T0094290

In the Land of Marvels

Also in the Series

In the Land of Marvels

Science, Fabricated Realities, and Industrial Espionage
in the Age of the Grand Tour

PAOLA BERTUCCI

Johns Hopkins University Press
Baltimore

This book has been brought to publication through the generous funding of the Frederick W. Hilles Publication Fund of Yale University.

Johns Hopkins University Press
2715 North Charles Street
Baltimore, Maryland 21218
www.press.jhu.edu

Cataloging-in-Publication Data is available from the Library of Congress.
A catalog record for this book is available from the British Library.

ISBN: 978-1-4214-4710-0 (hardcover)
ISBN: 978-1-4214-4711-7 (ebook)

Special discounts are available for bulk purchases of this book. For more information, please contact Special Sales at specialsales@jh.edu.

CONTENTS

The roots of this book date to almost two decades ago. I started my academic career in the United States soon after publishing *Viaggio nel paese delle meraviglie* in 2007. I thought then that my first project as an assistant professor would be an English version of it, but I soon came to realize that it did not make sense to do so while the tenure clock was ticking. It took several more years, a pandemic, and various stages of metamorphosis to see the project completed. It now feels as if I've come full circle.

The present book evolved from an initial idea to publish the transcription of Nollet's travel diary prefaced by a long introduction. I'm grateful to Greg Brown for his warm support of that project and to Zakiya Hanafi for providing me with an English translation of my Italian book. I am not sure how many sentences from her translation remain untouched in the present volume, but I couldn't even have started without an English draft to work on. Nollet's travel diary was transcribed decades ago by Judy Fox at Berkeley and was kindly given to me by John Heilbron and the late Roger Hahn. Abigail Fields created an electronic version of the diary, checking the Berkeley typescript against the original manuscript. Jennifer Strtak gave the transcription the final touches. Many thanks to them for this important work. The transcription of the diary can be found on the website of Johns Hopkins University Press at https://www.press.jhu.edu/books/title/12934/land-marvels#book_resources.

I'm enormously grateful to the archivists and librarians who provided photographic materials reproduced in this book, often after my last-minute, urgent requests. It is my pleasure to express my deepest thanks to Yasmin Ramadan (Beinecke Library, Yale), Alessandra Lenzi (Museo Galileo), Michele Righini (Biblioteca dell'Archiginnasio, Bologna), and Melissa Grafe and Chris Zollo (Medical Historical Library, Yale).

I am grateful to Ann Blair, Tony Grafton, and Earle Havens for their encouragement and support in all stages of this project, and to MJ Devaney and the Johns Hopkins University Press staff for their wonderful work on the manuscript. I've accumulated innumerable debts of gratitude in the years between *Viaggio* and *In the Land of Marvels* to colleagues, students, and friends who discussed related aspects with me. Thank you to all, at Yale, Bologna, Oxford, Paris, Stanford, and beyond. This book is dedicated to all the mentors I had the good luck of meeting along the way. Thank you for your help in making my unusual academic trajectory a reality.

In the Land of Marvels

Introduction

On a spring day in 1747, Carlo Antonio Donadoni, bishop of Sebenico, along with two monks and his personal physician, arrived at the Venetian villa of Gianfrancesco Pivati. Affected by long-term chiragra and podagra—forms of gout in his hands and feet—the seventy-five-year-old prelate had braved the journey in the hope of finding relief from the ailments that oppressively limited his movements. The podagra prevented him from walking without the support of the two monks, while the chiragra made it impossible for him to flex his hands, which remained sadly "curled up in mid-air." The bishop knew Pivati as the inventor of the "medicated tubes," an extraordinary electrical treatment that promised to heal the most persistent afflictions.

The visit delivered what it promised. Pivati filled a hollow glass cylinder with an anti-apoplectic substance, connected it to a hand-cranked machine, and, through a metallic conductor, started extracting sparks from the bishop. In the span of an instant, the monsignor opened both his hands, made a fist, and had so much strength that one of the monks "was soon compelled to beg him to let go, so forcefully was he gripping him." The bishop could walk by himself, clap his hands, stamp his feet, and was so amazed by the prodigy he had witnessed on himself "that he knew not whether he dreamed or was awake."[1]

The news of the bishop's extraordinary cure circulated quickly among Europe's reading public. Taken seriously by some and generating skepticism in others, it stirred an international controversy. Pivati, who was a member of the prestigious Bologna Academy of Sciences, wrote a private letter to Francesco Maria Zanotti, the Academy's secretary, in which he boasted of his invention by detailing several other accounts of prodigious cures. Thrilled that the inventor of such a portentous remedy should be a member of the Academy, Zanotti hastened to publish the letter. A few months later, Giambattista Bianchi, professor of anatomy at the University of Turin, and Giuseppe Veratti, lecturer of anatomy at the University of Bologna and another member of the local academy, confirmed the marvelous properties of the medicated tubes, adding their own to the list of extraordinary cures.

The authority of a patient such as the bishop, the confirmations by Bianchi and Veratti, and the support of a prestigious institution such as the Bologna Academy of Sciences boosted the credibility of the medicated tubes. The possibility that such a simple device could effect instantaneous cures stirred the interest of numerous experts in the two fields that the medicated tubes conjoined: medicine and electrical science. All over Europe, physicians, surgeons, and experimental physicists eagerly engaged in experiments with the medicated tubes, but more often than not they reported failure. In 1749, the abbé Jean-Antoine Nollet, a Paris-based celebrity in the area of experimental physics, announced that he would brave the arduous undertaking of crossing the Alps to see with his own eyes how the Italians applied the treatments. He presented his journey as an exploratory mission on behalf of the entire Republic of Letters, the community of the learned: experts along with amateurs would finally get to learn why these cures only seemed to work in Italy.

A member of the Paris Academy of Sciences, physics tutor to the dauphin of France, and celebrated author of texts on experimental physics, Nollet spent nine months in Italy and on his return wrote a report in which he firmly discredited the medicated tubes. He described his encounters with Pivati, Bianchi, and Veratti, as a philosophical duel in which he had engaged in the name of truth: the opposing parties challenged each other, fought to defend their position, and in the end truth prevailed. Nollet underlined the rashness of the Italians' conclusions and cautioned his readers to guard against the dangerous love of the marvelous to which the Italian electricians had fallen victim, a fascination that distracted natural philosophers from the pursuit of truth and thereby made them prone to credulity and self-deception. The experiments Nollet performed in Turin, he recounted, revealed the ineffectiveness of the medicated tubes, while in Venice, faced with the unrelenting questions of his opponent, Pivati finally admitted that, after an initial relief, the bishop of Sebenico had relapsed into his initial state.[2]

If Nollet's published account were the only source of information about his journey to Italy, it would just add color to widespread assumptions about eighteenth-century Italy, experimental philosophy, and the Enlightenment. Nollet's report implicitly highlighted the decline of Italian science after the golden age of Galileo Galilei and the Accademia del Cimento (Academy of Experimentation), reified the love of truth as an epistemic virtue of the enlightened philosopher, and presented the "love of the marvelous" as a misguiding passion from a previous era.[3] Yet several less public sources offer an entirely different perspective on the significance of this tour and on the published version of it that circulated all over Europe. Among them, Nollet's manuscript travel diary is the richest.[4] The diary and these other archival sources reveal that the controversy over the medicated tubes was only a minor concern for Nollet during his time in the Italian states.

Fabricated Realities

When I first laid my eyes on Nollet's travel diary, I did not expect that this document would change the course of my research as dramatically as it did. I had certainly hoped to find revealing behind-the-scenes descriptions of Nollet's encounters with his Italian counterparts, but page after page I was confronted with something that at the time I did not quite understand. The manuscript consists of 440 handwritten pages, most likely put together from now-lost first drafts, that offer daily descriptions of Nollet's activities, starting on May 26, 1749, when his journey began, and ending on November 3, 1749, when he got back to Paris. Over a hundred of these pages deal with a topic that Nollet never mentions in his published reports: the production of silk threads in the Italian states, mainly Piedmont. No other topic, including electricity, gets as much attention in the document. The diary makes it obvious that Nollet's journey to Italy was in fact a secret mission related to the silk industry, but it does not offer enough information to understand the bigger picture. Who sponsored the mission, and why? Did it have any effect? Why Nollet?

I traveled to Paris several times to examine documents in the Archives de l'Académie des sciences and the numerous folders related to the silk industry in the Archives nationales. The results were routinely disappointing. Yet I did find cursory evidence that both before and after the 1749 Italian trip, Nollet reported to the French Bureau of Commerce on matters related to silk. As I started learning more about the manufacture of silk threads, realizing its strategic importance in the eighteenth-century global economy, I also began to lose hope that I would ever find revelatory documents about Nollet's Italian journey in the archives.

Several years later, when I was again in Paris for unrelated reasons, I thought I would give the archives a final look. On the last day of my trip, after leafing through the usual mass of disappointing documents, during the archive's closing hour, while frantically flipping through the last sheets in the last folder, I found a small group of letters that Nollet secretly sent from Italy to his superiors at the Bureau. Sweating with excitement and fear that I would run out of time, I took photos. When I finally read the letters with due attention, a new area of research opened up for me. The letters cast new light on Nollet's journey to Italy and prompted me to think harder about the categories of industrial espionage and public science in the age of Enlightenment. I presented the result of my first exploration on Nollet's secret activities for the French state in a *Technology and Culture* article, which I subsequently developed into what now is chapter 1 of this book.[5]

During the COVID-19 pandemic, the proliferation of disinformation and misinformation on social media reminded me of the controversy over the medicated tubes and its extraordinarily long life on the printed page. As debates on vaccines and presumed miraculous cures for COVID-19 filled the digital sphere, and travel bans made my new, unrelated, research project hard to pursue, I revisited a past intention to publish an English version of my *Viaggio nel paese delle meraviglie* (A journey in the land of marvels). Instead of working on a translation, I decided to write a new book with a fresh argument. I realized that I could use my more recent archival discoveries to discuss the fabricated realities that experimenters like Nollet and Pivati created for their readers. Historians of early modern information systems have discussed primary sources created to intentionally deceive, especially with respect to political propaganda.[6] I show here that several eighteenth-century naturalists and experimental philosophers employed similar methods. While a central concern in the history of early modern science has been to investigate how communities of experimenters reached consensus on what counted as matters of fact, Nollet's journey to Italy offers the opportunity to reflect on the various ways in which the manipulation of information could affect scientific practice, discourse, and reputations.

The historiographical framework I first used in documenting the dispute between Nollet and his counterparts in Italy derived from a classic work in early modern science, Steven Shapin and Simon Schaffer's *Leviathan and the Air-Pump*. A study of a controversy between Robert Boyle and Thomas Hobbes, two major figures in the history of early modern natural philosophy, this foundational work approaches the dispute by "suspending judgment" on who was right and who was wrong, instead taking both sides seriously. Building on a key tenet of the sociology of scientific knowledge, the "symmetry principle," *Leviathan and the Air-Pump* offers a powerful model for studying scientific controversies, one that makes losers as relevant as winners and that shows that closure is the result of negotiations among historical actors rather than the unavoidable triumph of a self-evident truth.[7] This was also my approach: I was able to reconstruct the controversy both from the perspective of Nollet and of his rivals, which revealed that the official version emerged from behind-the-scenes negotiations among the parties involved. In doing so, I foregrounded the lesser-known context of electrical research in the various Italian states, which paved the way for the important contributions of later and more famous experimenters such as Luigi Galvani and Alessandro Volta. My 2007 book argues for the international relevance of Italian experimental culture in the eighteenth century by highlighting the reciprocity of the exchanges between Nollet and his counterparts.

What I did not realize at the time is that historians, just like readers in any time period, may fall prey to deceptions orchestrated by their own historical actors, even, or maybe especially, centuries after the fact. In this new book, I uncover the deceptions, fabrications, and manipulations in which the historical actors conspired, returning to the Pivati-Nollet controversy with the bolder argument that the philosophical duel existed only on the printed page. While Anglophone readers will still find the original contributions that my Italian book offered, here I provide fresh insights into the relationship between scientific expertise and information cultures in the eighteenth century. I take inspiration from the abundant, and often exhilarating, literature on early modern hoaxes, fakes, and forgeries, giving it a different spin. I argue that the printed page gave life to a philosophical duel that never was. My aim is not so much to label it a fake, a hoax, or a forgery but to cast light on the eighteenth-century analogue of the alternate realities created by information and misinformation circulating on the internet. I show that fabricated accounts had both a long life on the printed page and long-lasting effects: they boosted careers, fueled research, and contributed to the articulation of stereotypes and imagined pasts.

Stories that circulated on the printed page mediated the readers' perception of the real. This print-mediated reality overlapped in various degrees with factual reality. It could represent it accurately or it could be completely fabricated— just as twenty-first-century fake news reifies alternate realities in which events that never happened are perceived as real. In 1691, the bookseller John Dunton presented *The Athenian Mercury*, a journal he published as the periodical of the Athenian Society, a learned association that did not exist in real life. The Athenian Society, whose made-up history Dunton also published, soon earned numerous admirers, among whom was Jonathan Swift. As Adrian Johns comments, the Athenian Society "existed only in print, but appeared more real than the real thing."[8] In many other cases the relationship between factual and print-mediated reality was more nuanced: published versions of facts that did happen contained exaggerations, omissions, and other kinds of intentional manipulations. Embellished stories that circulated in print reified events in ways that differed from the experiences of those that lived them.[9]

Scholars of the early modern period have shown that fake news is not a recent phenomenon.[10] Printers and booksellers needed a constant flux of new content to keep their business afloat. They fed a variety of adulterated facts to readers, manufacturing news and unabashedly recycling old information that they repackaged as fresh. Depending on the nature of the event, their alterations consisted of a quick change of date, place, names of people involved, or other details that did not really alter the substance of the story or its appeal. These practices in the book trade owed their success to the reading public's fascination with the sensational and the prodigious. This fascination extended to the natural domain. Extraordinary phenomena occurring in nature along with prodigious cures were sure to bring printers good returns.[11] One might wonder whether the numerous marvels and portentous events that animated the learned elites in the early modern period might in fact be artifacts of these printing practices. But this would be another project. As I reexamined the primary sources related to the Pivati-Nollet controversy in light of this body of literature, I had in mind questions about the reliability of historical documents: How can historians be sure that their primary sources are not lying? How can they be confident that they are getting the inside jokes and not instead taking the fabricated realities as historical facts?

Several years ago, Ken Alder engaged in a practical joke that indirectly addressed these questions. He pretended to have discovered in an obscure Parisian archive a letter of the too-good-to-be-true kind authored by an eighteenth-century French forger full of verisimilar references to the experimental culture of the time. He published his pretend translation of the letter in a peer-reviewed journal. Without revealing his intentions anywhere in the article, Alder's aim was to expose the fine line between history writing and historically informed fiction.[12] Yet his hoax can also be taken as a challenge to move beyond the evidence offered by a single archive, especially if the documents in it seem to speak in one voice—and so in tune with one's expectations. Reconstructing from historical sources the practices leading to the fabrication of realities on the printed page may be quite difficult, or even impossible, for historians. This difficulty is in part what makes the controversy over the medicated tubes and Nollet's published account of his journey through Italy so compelling. The abundance of unpublished materials, both in France and in Italy, allows one to follow almost step by step the making of a philosophical duel that never was. Nollet's unpublished travel diary, kept in the Bibliothèque municipale de Soisson, in France, is one such source. I offer its transcription as an electronic appendix to this volume with the hope that broader access will encourage new studies, beyond what I present here and even beyond those that I can imagine.

Since the publication of Elizabeth Eisenstein's classic *The Press as an Agent of Change*, there has been a vast amount of scholarly discussion on print culture and its role in European transformations. Johns criticizes Eisenstein's technological determinism from the perspective of the history of science, calling attention to the multifaceted cultures of knowledge within which the print trade operated and that it helped create. In particular, Johns shows that printers and booksellers, as "manufacturers of credit," were essential in fostering scientific reputations and establishing matters of fact about the natural world.[13] Building on these insights, I demonstrate how the manufacture of credit enabled by the printed page could in fact result in the opposite of matters of fact: the creation of fabricated realities. When published, fake news, adulterated facts, or simply stories that authors embellished at their discretion and for their own purposes were difficult to distinguish from reality. I interrogate the functions that the fabricated stories served. I show that both parties that participated in the controversy over the medicated tubes concocted accounts whose published versions had very real effects both on their own careers and on the nascent science of electricity. I use the term "philosophical duel" to distinguish the fabricated accounts circulating in print from the historical reality of the controversy over the medicated tubes. By doing so, I show that the attempts to control the narrative on the printed page was one of the skills that defined the protagonists' careers.

Aware that a leading role in a sensational controversy over the most fashionable science of the century offered them tremendous publicity, both Pivati and Nollet carefully crafted what we could call their "print presence," the analogue of contemporary "digital presence": Nollet presenting himself as the foremost expert in electrical matters and Pivati advancing himself as the inventor of medical electricity. The emerging market for scientific news enabled fabricated experimental results to appear as matters of fact for readers all over Europe, and the ensuing debates fed interest in electricity and its medical applications. The fabricated story of the philosophical duel, in addition, served as a perfect cover for the secret service that Nollet performed for the French state.

Scholars like Robert Darnton for France and Marino Berengo and Mario Infelise, among others, for the Italian states, have brilliantly brought to life the active role of printers in the circulation of Enlightenment ideals, and the variegated world of hacks, literary pirates, and booksellers.[14] This book builds on this bulk of scholarship to show that reputed experimental philosophers too manipulated the printed page, with the goal to achieve publicity and reputation. The focus of my analysis, however, is not so much on print culture or the print trade, but on the role of the fabricated realities that naturalists and experimenters like Nollet concocted. Printers and booksellers do receive their share of attention in this book, particularly in the discussion of the processes through which Pivati acquired credit in the field of experimental philosophy, but my questions concern the envisioned effects of the artificial realities created on the printed page. I show that fabricated stories, just like the fake news that circulates in the digital world, had real-life consequences: they secured or started careers, promoted new medical treatments, and spread long-lasting stereotypes.

I also take inspiration from works in the history and sociology of scientific knowledge that have addressed manufactured scientific controversies and the fabrication of data in scientific literature. Scholars like Naomi Oreske, Robert Proctor, and Leah Ceccarelli have shown that by circulating verisimilar yet false or purposefully adulterated scientific information, twentieth-century corporation-friendly scientists took part in concerted campaigns aimed at eroding the credibility of widely accepted reports that could have damaged corporate interests, most famously related to climate change and the effects of tobacco on human health. These studies point to the power of fabricated data to insinuate uncertainty where there is consensus or to present closure where there still is controversy, documenting the real-life result of eroding trust in the scientific enterprise. Other studies have addressed instead scientists' intentional misrepresentation of their research owing to the pressures of career building (i.e., the tenure-track system) or conflicts of interest. Focusing on debunked articles published in peer-reviewed journals and subsequently retracted, these studies have shown that articles based on adulterated data continue to be cited in magazines or social media. That is, whether created within or outside the scientific community, scientific frauds do not quickly disappear: even after retraction or debunking, they continue to shape public debates.[15] These practices have a history that predates the digital era. Pivati's electrical cures and Nollet's philosophical duel were both published under the aegis of a scientific academy, respectively, in Paris and Bologna. The support of these prestigious institutions made it harder for readers to doubt the related facts. Pivati's "medicated tubes" survived Nollet's scathing attack, while philosophers writing decades later referred to Nollet's philosophical duel to present experimental physics as a tool of enlightenment and, paradoxically, a weapon against deception.

Although the vetting system that characterizes peer-reviewed articles was not in place in the eighteenth century, the possibility that even respectable savants could deceive colleagues along with the general public was a widespread concern. In 1715 Johann Burckhardt Mencke, the editor of the scientific journal *Acta eruditorum*, published *De charlataneria eruditorum*, an influential work that was soon translated into English as *The Charlatanry of the Learned*. Mencke's work, whose frontispiece features the ancient motto "mundus vult decipi" ("the world wants to be deceived"), argues that deception is by no means the province of quacks and mountebanks and compares scholars and naturalists to actors in search of an audience who do not hesitate to deceive their readers in order to boost their reputations. *The Charlatanry of the Learned* generated a debate that turned the ability to unmask deception and even to publicly shame frauds into a distinctive feature of the honorable scholar.[16] The echoes of this debate reverberated in Nollet's published account of his journey. He presented the Italians' love of the marvelous as a form of self-deception that a lover of truth like himself felt obligated to fight on behalf of the entire community of the learned. He quoted the same *mundus vult decipi* motto to dismiss those who did not believe his version of the story.[17]

Nollet's published account of his journey through Italy offers a counterpoint to the "anti-marvelous" ethos of the Enlightenment articulated by Lorraine Daston and Katherine Park in their classic *Wonders and the Order of Nature*. Daston and Park argue that enlightened elites defined the love of the marvelous as a passion of the vulgar. They observe that over the course of the eighteenth century wonders and marvels disappeared from the vocabulary and sensibility of the Republic of Letters because enlightened intellectuals simply ignored them—without engaging in any crusade or duel.[18] In contrast, I show that Nollet's philosophical duel was predicated on the idea that the love of the marvelous could and did deceive the learned along with the vulgar. His account highlighted that Italian elites, not just the vulgar, had fallen prey to this dangerous threat that could only be neutralized with the weapons of experimental philosophy. I argue that for authors like Nollet, exposing deceptions and self-deceptions—whether in print or in practice—was equivalent to pursuing intellectual enlightenment, in the Kantian sense of emancipation from a self-imposed state of immaturity.

From the controversy over the medicated tubes, I zoom out to the multifaceted culture of travel to Italy, bringing scholarship in the history of science in conversation with the vast historiography on the Grand Tour. The Grand Tour and its armchair-traveler version is a dominant paradigm through which eighteenth-century travel to Italy has been conceptualized. Critique of this paradigm has shifted the focus from Italy to northern European countries and to the reverse Grand Tour: travels that Italians undertook north of the Alps.[19] Yet, even in the age of the Grand Tour, Italy was a destination for other kinds of travel. Nollet was one of many travelers who toured Italy in search of the "secret" of its silk threads. In addition to his secret mission, Nollet presented various accounts of Italian technology to the Paris Academy of Sciences, addressing methods for making bricks and mortar, the construction techniques for houses and other buildings, and the creation of frescoes and tiles for decorating interiors. He also described where and how Italians grew hemp, maize, rye, and wheat, illustrated rice paddies, and introduced his fellow academicians to chickpeas and polenta. The full report of his Italian journey took no less than twelve sessions of the Academy to read over the course of a year and was published in the *Mémoires de l'Académie Royale des Sciences* in two long articles that amount to a total of ninety-eight pages, of which only sixteen are dedicated to the medicated tubes. Nollet's own choice of topics, then, points to the French public's interest in Italian technology and industry.[20] I argue that Nollet's published versions of his journey relied on stereotypes that travel literature circulated for centuries, which represented Italy as a land of marvels and the Italians as lovers of the marvelous. The stereotypes provided the foundations for his fabrication of a duel against the love of the marvelous that affected Italians.

Land of Marvels

The journey to Italy started with a book. Sometimes it continued by carriage but more often in the reading room. In the age of the Grand Tour, travelogues about Italy constituted a best-selling genre. Catering especially to a growing public of armchair travelers, these publications turned stationary readers into seasoned connoisseurs, even when they had never entertained the idea of undertaking the fatiguing endeavor of actually crossing the Alps. In the light of an oil lamp, armchair travelers did not need a passport or letters of credit, nor did they have to contend with the hardships of traveling by horse-drawn carriage, the jolts caused by rough roads, the risk of attacks or epidemics along the way, or the endless inspections by customs officials. Only in their imagination would they have to climb in and out of their carriage multiple times during the two days needed to cross the Alps, complete long and strenuous stretches on foot, horseback, or sled, and watch the carriage itself being dismantled and reassembled, amid snowstorms and icy winds. By simply turning the pages of one of the many available travel guides, armchair tourists could go from Venice to Rome literally in the blink of an eye. It was thanks to travel accounts that reading and talking about Italy became an opportunity for exhibiting refinement and erudition. Conversing about artistic masterpieces and the Italian landscape, showing interest in classical antiquity, being able to appreciate ironic quips on the customs and traditions of the Italians was tantamount to vaunting culture and sophistication.[21]

Real-life travelers, just as those in their armchairs, could explore the peninsula following in the tracks of Michel de Montaigne or Richard Lessels, compare the opinions of Joseph Addison with those of Anne-Marie du Boccage, or contrast John Evelyn's Italy to that of Jérôme de Lalande. On their return, they contributed to the long list of publications on the topic with their new take on Italian marvels. The "voyage d'Italie" was a well-established literary genre that boomed during the eighteenth century. Gilles Boucher de la Richarderie's bibliography of travel books, *Bibliothèque universelle des voyages* (Universal library of travels), which was published at the end of the eighteenth century, listed about thirty titles on Italy up to 1700 and over seventy new ones published between 1700 and 1799. Travel literature, along with salon conversations, reports sent by academicians *en tour*, paintings, engravings, and news published in the gazettes constructed "Italy," which did not exist as a political entity, as a single country. Travelers passed through a constellation of several states differently ruled, each with its own currency and even local language, yet their accounts did not generally linger on this fragmentation (fig. I.1). They represented Italy as an open-air *Wunderkammer*, overflowing with magnificent works of art, extraordinary natural phenomena, and unconventional social customs.

Travel accounts of Italy also described a natural world that surprised and intrigued. Destinations such as Vesuvius, Acqua Zolfa, Pietramala, the Cave of the Dog, and other unusual natural formations punctuated the itineraries of grand tourists just like the Colosseum or the Uffizi Gallery. Even Nollet employed the vocabulary of the marvelous when he saw nocturnal flashes in the Venetian lagoon: in the evening nearby the gondolas, he explained, close to the walls of the houses and wherever an obstacle forced the water to swell, it emitted very bright sparks and glimmered "in a marvelous way." This

Fig. I.1. The Italian states around 1740. Map drawn by Johann B. Homann for the 1752 Homann Heirs' *Maior atlas scholasticus ex triginta sex generalibus et specialibus.* Beinecke Rare Book and Manuscript Library, Yale University, New Haven, CT.

extraordinary spectacle recurred almost every time the gondoliers hit the water with their oars, giving rise to a series of "fiery tongues" lighting up from the depths of the lagoon.[22]

Publications detailing Italy's unusual natural phenomena turned remote locations into conversation topics and new travel destinations. They created imagined realities that real-life travelers could confirm or debunk. Learned tourists eagerly traveled to Pietramala, a small village in the Appenines on the road to Florence, to eyewitness the famous "fires" described by Athanasius Kircher in his 1665 *Mundus subterraneus* (The subterranean world), an omnipresent reference for travelers interested in natural phenomena. The "fires of Pietramala" were myriad flames of various size and colors that sparked from the rocks, especially spectacular at night. In 1706, the astronomer Francesco

Bianchini guided a group of "curious" foreigners to Pietramala and recorded their observations, which he then sent to the Paris Academy of Sciences. Calling the phenomenon "a marvelous effect of nature," he invited other naturalists to dedicate their attention to it.[23] Pietramala thus became a site for observation and experimentation that attracted naturalists and tourists alike. Publications that grew out of these trips amplified general interest.[24]

While they admired Italian nature, travelers compared what stood in front of their eyes with the imagined realities created by travel accounts. Maximilien Misson, the author of the bestseller *Nouveau voyage d'Italie* (*A New Voyage to Italy*), found the phenomenon even more marvelous than Kircher had described: he saw a tall fire that the rain seemed to extinguish only to rekindle it, making it more vigorous than before.[25] On the contrary, Charles De Brosses, president of the Parlement de Dijon, who toured Italy in 1749 and had read both Kircher's and Misson's accounts, flagged Misson's description as an exaggeration and offered his own interpretation based on new studies of phosphorescence. For him, the rocks did not conceal any mystery: like phosphorus, they absorbed and released natural light.[26]

Misson's *New Voyage to Italy*, a book in the form of several letters the author wrote during his journey through Italy in 1687 and 1688, was one of the most influential travel guides of the time. Published for the first time in 1691, translated into English four years later and soon afterwards into Dutch (1704) and German (1713), the book went through numerous updated editions until the early 1800s and was a model for other travel books, including Joseph Addison's celebrated *Remarks upon Several Parts of Italy* (1710). In line with the tradition of the *ars apodemica*, travel advice literature in vogue since the sixteenth century, Misson discussed Italy's monuments and gave practical information for undertaking the journey with peace of mind. Like other travel accounts, the book normalized the fragmentation of Italy into various states as one of the bizarre features of the Italian wonderland.[27] Misson also described marvelous animals and natural sites, introducing readers to dragon-like sea horses (fig. I.2) and dangerous tarantulas whose poisonous bites could only be cured by music and dance.[28]

As they described the Italian marvels, travel accounts like Misson's othered Italy and its inhabitants from the rest of Europe, perpetuating stereotypes that had originated much earlier and that became a filter through which travelers conceived of Italy and its inhabitants.[29] Prone to exaggeration, infatuated with their country beyond measure, and abusers of superlatives, Italians were lovers of the marvelous whose "pompous and bombastic words" often deceived.

Fig. I.2. An illustration of a seahorse found on the sand in Fano, Italy, in Maximilien Misson's *Nouveau Voyage d'Italie* (1722). Notice how the background and exaggerated features make it appear gigantic and dragon-like. Creative Commons, Université de Poitiers, Poitiers.

Tourists had to beware of Italians' tendency to exaggerate: "We have already seen I do not know how many presumed eighth wonders of the world," Misson noted sardonically. The Italians' way of talking, he observed, which ranged from soft and playful to childish and hyperbolic, was dangerously charming. Their lack of judgment was like a contagious disease that infected foreign travelers. Back in their countries, tourists were often eager to tell "great stories about very little things."[30] Tourists prepared for their journey by reading as many travel guides as they could find, absorbing information and stereotypes. While De Brosses often disagreed with Misson, he shared his opinion regarding Italians' poor judgment: "By God! The Italians squander on superlatives. It costs them nothing, but it costs a lot to foreigners, who spend a lot of money and time to see things that are highly praised, yet unworthy of being so."[31]

This construction of Italians as unreliable lovers of the marvelous was a literary commonplace that travel literature consolidated over time through multiple examples. "Leaning in a frightful way," as Montesquieu remarked, the tower of Pisa was for some travelers a clear indication of the Italians' bizarre nature.[32] As Misson recounted, some observers believed that the leaning tower was the intentional design of an eccentric architect. In support of this view, authors mentioned other Italian towers that sported a certain "leaning air." The Garisenda tower in Bologna, for example, allegedly offered evidence that the Italians had a peculiar "taste for these off-perpendicular constructions."[33] De Brosses mentioned that according to some travelers, the Garisenda tower in Bologna was intentionally built off the perpendicular "out of malice" or "to scare passersby, who are given the impression of seeing it come down on them."[34] According to Nollet, it was not a "national taste" for bizarre architecture that accounted for the unusual way the Garisenda was built, as others maintained; rather, the "peculiar fantasies" inspired by the leaning tower of Pisa, a monument that was "exceedingly marvelous to the eyes," did (fig. I.3).[35]

Italians living abroad sought to counter the caricatures of Italian social habits, civic and religious rituals, and artistic tastes that travel literature spread. For example, in 1767, Giuseppe Baretti, an Italian living in London, published *An Account of the Manners and Customs of Italy with Observations on the Mistakes of Some Travelers with regard to That Country*, a scathing assessment of travelogues produced by foreigners. As a journalist, he was aware of the power of the printed page to circulate embellished stories that went on to enjoy a life of their own. He was particularly concerned with the biases ensuing from travel accounts' fabrications. While the goal of traveling purportedly was to expand one's cultural horizons, Baretti pointed out that numerous accounts in fact spread false information and reinforced national prejudices: "Thus falsehood is palmed for truth for the credulous and thus are men confirmed in a narrow way of thinking and in those local prejudices, of which it ought to be the great end of travelling, and books of travel, to cure them."[36]

Works on cross-cultural encounters have shown that travel narratives articulated epistemological hierarchies and conceptualized ways of framing

Fig. I.3. The Garisenda tower as represented in Maximilien Misson's *Nouveau Voyage d'Italie* (1722). This is a fantastical representation of the tower, as the lower structure in the image is in fact absent. The tower, which dates from the twelfth century, began to lean early on. Its appearance inspired Dante, who mentions it in the *Inferno* (canto 31). Beinecke Rare Book and Manuscript Library, Yale University, New Haven, CT.

non-European people as primitive.[37] Within Europe, the representation of the Italians as lovers of the marvelous could achieve similar effects. As Melissa Calaresu has shown, however, while travel narratives doubtless "exoticized Italians," Italian learned elites participated in the same culture as foreign travelers and answered foreigners' stereotypes about them with their own stereotypes about foreigners.[38] To further advance this important point, I contrast the imagined realities created by travel accounts with the real-life encounters between Nollet and the local communities as they emerge from private documents. My analysis shows that while travel accounts continued to spread stereotypes about Italy and the Italians, the real-life exchanges between travelers and their Italian hosts typically unfolded in a culture of reciprocity: they exchanged information, gifts, membership in academies—and they also reciprocated stereotypes.

Intelligent Travel

Travel was quintessential to the culture of Enlightenment. It supported the Republic of Letters, enlightened economies, and colonial machines.[39] In his *Sentimental Journey*, Laurence Sterne satirized the increasing heterogeneity of eighteenth-century travelers with a taxonomy that included "Idle Travellers, Inquisitive Travellers, Lying Travellers, Proud Travellers, Vain Travellers, Splenetic Travellers, . . . the Travellers of Necessity, [t]he Delinquent and Felonious Traveller, [t]he Unfortunate and Innocent Traveller, [t]he Simple Traveller," and "last of all (if you please), [t]he Sentimental Traveller."[40]

This book presents Nollet as a type of traveler that escaped Sterne's attention: the "intelligent traveler." At a time of great industrial innovations, travels like Nollet's, aimed at gathering intelligence on technical matters, unfolded along the same routes followed by grand tourists and learned savants. These journeys are often grouped together under the anachronistic label of industrial espionage. I propose the alternative notion of intelligent travel to examine more carefully the strategies that presumed "industrial spies" adopted to fulfill their task. Secrecy in particular, as one of these strategies, should not be taken as a monolithic term whose meaning remained unchanged through time.

Eighteenth-century readers were accustomed to the idea that travelers could be spies, but they were unfamiliar with the notion of the industrial spy. The word "espionnage"—"the action of spying, the craft of the spy"—made its debut in the *Dictionnaire de l'Académie françoise* only at the turn of the nineteenth century.[41] Yet it was a recurring neologism in French popular epistolary novels, such as Jean Paul Marana's *L'espion turc* (*Letters Writ by a Turkish Spy*), Mathieu-François Pidansat de Mairobert's *L'espion anglais* (The English spy), Ange Goudar's *L'espion francois à Londres* (*The French Spy in London*), which Goudar presented as the French counterpart to *The Spectator*, and Goudar's *L'espion chinois* (*The Chinese Spy*).[42] These texts catered to the readership's fascination with foreign news, travel literature, satire, and secrecy, offering

fictional, often satirical, stories of individual travelers who reported diplomatic news or courtly gossip from a foreign country. In a captivating twist
on the *ars apodemica*, these novels linked the observation of people, events,
arts, and customs with "spying." The choice of the term "spy" in these titles
was a satirical one, yet the actions of spying and observing are also associated in the etymological root of the verb "to spy," from the Latin for "to
observe," and in the materiality of the telescope, initially called "spyglass."[43]
The fictional observers in these "spy" stories, however, did not focus on technical knowledge. Just like their contemporary real-life spies, these fictional
spies operated within strategic domains commonly controlled by the state;
they participated in diplomatic missions, took part in military strategies, and
collaborated with the police and, in the case of Catholic countries, with the
Inquisition.

In the age of Enlightenment, the secrecy that accompanied the absolutist
state created by Louis XIV in concert with Colbert was a subject of critical examination and the morality of spies was under public scrutiny.[44] In his *Esprit
des lois* (*Spirit of the Laws*), Montesquieu advised good monarchs not to employ spies, famously declaring that "espionage might perhaps be tolerable if it
could be exercised by honest people." Instead, spies were abject men, and their
ignominy reflected on the reputation of the king himself.[45] Intellectuals such
as Diderot and Voltaire also publicly stigmatized secrecy.[46] The *Encyclopédie*
article "Spy," however, explained that "good spies" were essential to the military art and that the state should spare no cost for maintaining them. Spies
were particularly devoted state servants, who risked their lives for the crown.[47]
Similarly, the anonymous *Traité des ambassades et des ambassadeurs* (Treatise
on embassies and ambassadors) highlighted the value of secrecy, remarking
that "a man who speaks all he thinks, and who has not aged in the habit of
never revealing his secrets, is unable to manage state affairs."[48]

Goudar, who was an outspoken critic of the French political system, made
the increased amount of French secret activities the target of his satire. In *The
French Spy in London*, he feigned criticism of Montesquieu's negative assessment of the spies' morality, explaining that a multitude of "honest people" had
engaged in espionage, with remarkable social returns: "The prince's friend and
the spy have become very consequential men: the former makes the pleasures
of the Court, the latter those of the City."[49] In Goudar's work, every traveler
who reported information from one country to another was a spy of some sort;
spies were everywhere, and their ever-increasing number led to specialization.
Paraphrasing Sterne, Goudar offered a satirical taxonomy of the various spies

traveling across the globe: there was "the spy of the court, the spy of the city, the spy of the palace, the spy of the table, the spy of the bed, the spy of the street, the spy of the game, the spy of men, the spy of women, the spy of spectacles, etc."[50] Among this vast constellation of spies, however, there was no industrial spy.

This absence is even more glaring when we consider that Goudar himself gathered technical intelligence for the French state in his youth, passing reports on the manufacture of londrins, a type of fabric, to the French ambassador in Constantinople. Other members of his family undertook similar intelligent travels, including his father, Simon Goudar, in his official capacity as the general inspector of manufactures in Languedoc, and his brother François, the owner of a textile manufacture, who traveled to Turkey in the 1740s to "find the secret of dying cotton in red, in the style of Adrianople."[51]

The category of industrial espionage has been indelibly applied to the eighteenth century by John Harris, whose capacious *Industrial Espionage and Technology Transfer: France and Britain in the Eighteenth Century* provides compelling evidence that trade and technological developments were driving factors for eighteenth-century travel along with cultural exchange and individual refinement. Harris focused on the French attempts to import British technology, yet the geography of industrial travel was far from unidirectional.[52] Both Britain and France, the two imperial powers at the core of Harris's book, repeatedly attempted to import silk technology from the kingdom of Piedmont. In the course of the eighteenth century, individual travelers and state-sponsored missions targeted small villages in Piedmont's countryside in the attempt to challenge the kingdom's primacy in the production of silk threads.[53] Nollet, as I show, participated in these efforts.

Was Nollet a spy? The answer to this question calls for a careful examination of the persistence of earlier practices of secrecy and dissimulation in the age of public science. The public culture of science has long been the main lens through which we have understood the Enlightenment's engagement with scientific knowledge and, more recently, the early processes of industrialization and the Enlightenment economy.[54] The emphasis on the public, however, has obscured the role that secrecy continued to play in the eighteenth century.[55] Notably, the eighteenth century constitutes a gap in the "State of Secrecy" issue of the *British Journal of the History of Science*, which explores the relationship between science and secrecy over several ages.[56] Nollet's secret mission offers an opportunity to enter this unexplored territory. In particular, it shows that practices of secrecy and dissimulation, typical of earlier periods, persisted well into the eighteenth century.[57]

I label Nollet an intelligent traveler rather than an industrial spy to illuminate this legacy from earlier times and to foreground the collective nature of technical information gathering. The category of industrial espionage reinforces notions of technology transfer that center on individuals and machines, neglecting more complex sociolegal dynamics that often played a more relevant role. It also bestows a romantic and simplistic aura of secrecy on journeys, like Nollet's, that were not entirely secret. Harris himself had to acknowledge that there were "varying degrees of espionage in different tours devoted to intelligence gathering," ranging from "a tour apparently merely seeking education and enlightenment," which he branded "innocent espionage," to "an expedition of straightforward spying."[58] As Montesquieu and other authors pointed out, however, a spy was never "innocent." Those who engaged in espionage were well aware of what they were doing, for whom, and of the risks they were taking. Spying required planning, and naïveté was risky. If we substitute the word "espionage" with "secrecy" in Harris's sentence, we get a more accurate picture: there were indeed varying degrees of secrecy in the intelligent travels that the French state sponsored throughout the eighteenth century.

Intelligent travels were sometimes secret, but many other times they were not. Often, they were somewhere in between. Inspectors of manufactures engaged in intelligent travel in fulfillment of their job. Merchants, artisans, or savants, instead, often resorted to some form of secrecy to ensure the success of their missions of technical intelligence gathering. A notable example of an "open" intelligent travel is the three-volume *Voyages métallurgiques* (Metallurgical journeys) by France's general inspector of manufactures, Gabriel Jars the younger. Published in 1775, it consists of numerous reports on mining that Jars compiled during his journeys in various European countries.

Several scholars have discussed the changing meaning of secrecy in history. Pamela Long's pioneering work on the mechanical arts from antiquity to the sixteenth century has brought to light a wealth of sources that indicate that the historical relationship between openness and secrecy was much more complex than a polarized opposition. She demonstrates in particular that artisans could simultaneously advocate openness and protect secrecy.[59] Other historians of science and medicine have contributed to further blurring the line between openness and secrecy until the turn of the eighteenth century.[60] Nollet's Italian journey illuminates the continuities into the eighteenth century of trends that these scholars have so incisively discussed. I uncover the multifaceted culture of secrecy in which his intelligent travel unfolded, arguing that Nollet's reputation as a man of science served as an effective cover for his secret mission and that, in turn, his secret service boosted his career.

Structure of the Book

Nollet's secret mission forms the bulk of chapter 1, "Silk and Secrets," which examines what made him an ideal candidate for gathering information on silk manufacture in Italy. I argue that Nollet's public reputation, along with his technical skills, made him an ideal candidate for the operation of technical intelligence gathering that the French Bureau of Commerce needed. The chapter shows that Italian informants were fundamental players in Nollet's scheme and demonstrates that earlier practices of secrecy were alive and well in the eighteenth century. By calling attention to the international importance of Italian silk technology, the chapter also highlights the relevance of Italy as a travel destination beyond the paradigm of the Grand Tour.

Chapter 2, "Electricity, Enlightenment, and Deception," focuses on the nascent science of electricity in Europe, analyzing the context in which the controversy over Pivati's medicated tubes unfolded. I argue that the numerous publications on electricity created an imagined reality in which spectacular electrical demonstrations could be easily replicated, yet it was through the activity of several itinerant demonstrators that a vast range of Italian amateurs learned how to perform experiments. I show that the medical applications of electricity emerged from these spectacular performances and were perceived by many as dangerously close to the kinds of fake remedies offered by medical charlatans. The chapter offers a discussion of Nollet's own contribution to medical electricity and demonstrates that lack of consensus on the curative power of electricity, and particularly the controversial nature of electric cures, generated debates that—once in print—fueled the popularity of the new science.

Chapter 3, "Fabricated Controversy," discusses the making of the philosophical duel that never was. It situates Pivati's experiments with the medicated tubes in the context of the Venetian book trade and early eighteenth-century encyclopedias. Pivati was the superintendent of the book trade in Venice and the author of the first encyclopedic dictionary in Italian, *Nuovo dizionario scientifico e curioso, sacro-profano* (New dictionary, scientific and curious, sacred-secular), published between 1746 and 1751. The dictionary was a ten-volume work and a risky financial investment through which Pivati made both friends and enemies. I argue that Pivati fabricated news of his electric cures with the intention of promoting his dictionary in the various Italian states. Relying on unpublished documents, such as Nollet's travel diary and the correspondence among Nollet's Italian counterparts, I show that no parties involved in the controversy were ready to change their minds as a result of a confrontation based on the replication of experiments. The chapter discusses the fabrication of Nollet's published version and the erasures that it operated. It also examines the long life the philosophical duel enjoyed on the printed page and the reasons it remained so appealing to later generations.

Chapter 4, "Natura Marvels, Instruments, and Stereotypes," focuses on the role of natural history in consolidating stereotypes about Italians. It examines how naturalistic writings contributed to the construction of Italy as a land of marvels, highlighting the role of Italian authors and touristic guides in such construction. The chapter also considers Nollet's description of one of the most spectacular events in Naples: the miracle of the liquefaction of St. Januarius's blood. Nollet's account shows that, like other travelers, he was more fascinated by the Neapolitan crowd and the spectacle of their devotion than by the miracle itself. This spectacle epitomized the various ways in which deception played out in Italian culture.

The focus on the Italian scientific community that this book provides casts light on a heterogeneous culture of the eighteenth-century Republic of Letters, whose geography was more opportunistic than it appears from the traditional focus on publications, institutions, and metropoles. If the Grand Tour centers Italian art and ancient history as the main reasons for crossing the Alps, Nollet's intelligent travel offers quite a different perspective on the lure of Italy in the eighteenth century. Not Rome, but various small villages in Piedmont's countryside were the centers of attraction for travelers interested in technology and industry. Well before the age of the Watts and Wedgwoods, the various silk manufactures in Piedmont fascinated enlightened travelers, who admired the combination of ingenuity and efficiency embodied in productive machines.[61] Travelers' comments on the natural and cultural landscapes they encountered contributed to drawing the map of the Republic of Letters, marking its centers and peripheries. This book shows that such map reveals just as it conceals and distorts.

CHAPTER ONE

Silk and Secrets

On June 30, 1749, the secretary to the French intendant of commerce, Antoine-Louis Rouillé de Jouy, forwarded the first letter that the abbé Nollet sent him from Turin to Daniel Trudaine, the head of the French Bureau of Commerce. Declaring himself pleased with Nollet's diligence, the secretary reminded Trudaine of the purpose of Nollet's journey: the abbé would "provide necessary instructions to confirm the information we already have, or that we think we have" on the manufacture of silk threads in Piedmont, and he would help reorganize all available data according to "the most certain rules of physics."[1] This aspect of Nollet's journey remained secret until well after his death. Numerous publications, including Nollet's own, celebrated his Italian journey as a mission on behalf of the Republic of Letters to ascertain the claims of a group of Italian electricians about the prodigious medical properties of electrified tubes. This chapter examines why the Bureau entrusted this delicate operation to Nollet. It argues that Nollet's popularity in the field of experimental philosophy offered a perfect cover for a secret mission on behalf of the French state and shows the extent to which the paradigm of the Grand Tour obscures aspects of eighteenth-century Italian science and technology that made the lands south of the Alps an alluring travel destination.

The abbé Nollet

On his arrival in Italy, the forty-eight-year-old abbé Nollet was an international celebrity. A member of the Paris Academy of Sciences, of the Royal Society of London, of the Bologna Academy of Sciences, an acclaimed author of texts on experimental physics, and the owner of a successful workshop that produced scientific instruments, Nollet occupied a leading position in the Republic of Letters. Born on November 19, 1700, at Pimprez, a small village in the Picardy region not far from Paris, Nollet came from a modest family who had nudged him toward a career in the church. After obtaining a degree in theology, completed in Paris in 1724, he was ordained deacon but went no higher in the church hierarchy. The title of abbé, not to be confused with that of abbot, was commonly used for people who took the minor orders.

Immediately after graduation, Nollet began to work as a tutor for the Taitbout family, officers at the Hôtel de Ville in Paris. Around that time he also started an informal apprenticeship with the court enameller, Jean Raux, whose workshop was nearby and whose creations delighted, among many others, the royal children and Madame de Pompadour.[2]

Raux and Nollet were both members of the Société des Arts, an association that brought together a group of individuals from different social backgrounds, who shared an ambitious vision for the role of the mechanical arts in the commercial and colonial projects of the French state. One key tenet of the organization was that the ruling elites needed to be conversant in technical matters, an ideal that was at the core of the school of experimental physics Nollet established years later.[3] Upon admission to it in 1728, Nollet presented a terrestrial globe that he dedicated to the duchess of Maine, which revealed his ambitions to weave connections with the aristocracy. The globe included the itineraries of famous explorers and the dedication was inscribed in a cartouche with the personifications of the four continents, a clear hint at the relevance of the art of navigation to colonial pursuits.[4] Although very scant records of Nollet's activities in the Société survive, it is likely that his membership acted as a springboard for establishing connections with members of the Paris Academy of Sciences. In Paris, enamellers made glass instruments such as thermometers and barometers, along with the relevant components of pneumatic pumps, and electrical machines.[5] Nollet's skills in this art brought him to the attention of Charles de Chisternay Dufay, a member of the Academy whose most recent work concerned the mercurial glow visible under certain circumstances in barometric tubes. Soon afterward, René Antoine Ferchault de Réaumur, Dufay's patron within the Academy, entrusted Nollet with the direction of his laboratory and worked with him on thermometry and natural history.[6] After Nollet's election to the Academy in 1739, Réaumur assigned him the task of compiling a volume on the art of enameling, glassmaking, and glazing for the series *Descriptions des arts et métiers.*[7]

Nollet collaborated with Dufay on the promotion of experimental philosophy in France. In 1734, and then again in 1736, the two traveled to England and the Netherlands on state-sponsored trips, in which Nollet was charged with weaving connections with famous demonstrators. He interacted with John Teophilus Desaguliers and Pieter van Musschenbroek, two of the most famous experimental physicists of the time, who employed instruments specifically made for lectures open to paying audiences.[8] Nollet worked with them hands-on to perfect the techniques for making this kind of scientific instruments.[9] During his visit to England, he was elected a member of the Royal Society of London. On his return to Paris, Nollet started a successful business as an instrument maker. He trained and mentored artisans who later earned excellent reputations. One of them, Monsieur Cousin, was a guest of Voltaire and the marquise du Châtelet at their residence in Cirey. Voltaire, who believed that "all of Leibniz's theodicy was not worth a single experiment by Nollet," was impressed by Cousin: in addition to being a machinist and designer, noted Voltaire in a letter, "he studies mathematics, applies himself to experiments, and will shortly learn how to operate the observatory."[10] Voltaire purchased a fully equipped physics cabinet from Nollet at the exorbitant cost of ten thousand livres—the annual salary of a member of the Academy amounted to six thousand livres—and made a large down payment to ensure that the project got under way. To those who pointed out the risks of paying in advance, Voltaire replied that Nollet was not a common man: "He is a philosopher, a man of true worth, the only one who can fit out my physics cabinet."[11]

In 1734, Nollet opened a school of experimental physics in his home on rue du Mouton, near the Hôtel de Ville. In line with the program of the Société des Arts, the school's goal was to introduce the French public, in particular the elites, to the methods of the mechanical arts.[12] In his lectures, Nollet discussed topics related to "the practices of the arts and to the machines most commonly used for the convenience of civic life."[13] At the end of the course, his students would be able to more confidently assess inventions, invest in manufacturing, and promote building or mining projects. Nollet's educational program addressed the young in particular. Knowledge of machines and experimental apparatus would enable children to avoid "popular mistakes," "ridiculous beliefs," "false marvels," and other charlatanries that beguiled the wealthy and poor alike.[14] Nollet's combination of useful knowledge and spectacular demonstrations proved enormously popular among the French aristocracy. The marquise du Châtelet reported that Nollet's lectures had become à la mode, with "carriages of duchesses, peers, and pretty ladies" lined up in front of the school's entrance.[15] To guarantee a place in the school on rue du Mouton one had to book ahead. Nollet's instruments were likewise in demand and sought-after like the latest fads. The duke of Penthièvre (nephew of Louis XIV) and the duke of Chartres—both thirteen-year-olds—attended Nollet's lectures. Foreign visitors seeking the most fashionable events in the city too flocked to his lectures.

Nollet's celebrity reached well beyond Paris. In 1739, the king of Piedmont-Sardinia, Carlo Emanuele III, invited him to Turin to teach a physics course to the young prince, the duke of Savoy and future king, Vittorio Amedeo III. The king had acted on the suggestion of the prince's tutor, the marquis de Fleury, who had identified Nollet as a savant who combined proficiency on "ingenious machines" and perfect manners.[16] Nollet arrived in Turin at the end of May 1739 and spent six months there, during which he also taught a course at the university that attracted "at least two hundred people of all ages, sexes and social conditions."[17] Cadets from the royal schools in artillery and fortifications also flocked to Nollet's lectures.[18] The success of the demonstrations thrilled the king, who compensated Nollet with a sum equal to the annual salary of the local professor of experimental physics at the university.[19] The duke of Savoy expressed his gratitude to Nollet with a diamond ring.[20] Nollet's stay in Turin had important consequences for the teaching of experimental physics in the city, as the king donated all the instruments that Nollet brought for his lectures to the university.[21] The machines were incorporated into the physics cabinet (the collection of physics instruments), which in that very period was being reorganized as part of a larger project led by Nollet's future rival, Giambattista Bianchi, chief physician to the court and professor of anatomy. A bas-relief of Nollet was placed in the cabinet.[22]

Soon after Nollet's return to France, the Bordeaux Academy of Sciences commissioned an entire physics cabinet from him and invited him to offer public courses, which proved extremely popular.[23] While in Bordeaux, Nollet attended the salon of Madame Duplessis, widow of a *conseiller au Parlement*, who was, alongside her daughter, a famous patron of the sciences. Among the regulars of the Duplessis salon was Montesquieu, who did not particularly like Nollet, whom he regarded as nothing more than "Réaumur's lackey."[24] Montesquieu's son, Jean-Baptiste de Secondat, inherited this hostility, vehemently attacking Nollet's electrical theories years later.

By the time of his 1749 journey to Italy, Nollet's reputation was well established. In 1744, the king of France, Louis XV, invited him to court to teach experimental physics to the dauphin. Nollet moved to Versailles, bringing instruments that made an impression for the "taste and intelligence" with which they were crafted.[25] Many people at Versailles attended his lectures, including the queen Mary Leczinska and the wife of the dauphin, who demanded a personal course.[26] In addition to generous compensation, Nollet received further recognition from the king: in 1746, he was assigned the apartment of the deceased sculptor Guillaume Coustou I in the Galleries of the Louvre, where, ever since 1608, the court had housed the best artists and most skillful artisans.[27]

The Paris Academy of Sciences was closely tied to the French Bureau of Commerce.[28] Nollet's mentors at the Academy, Réaumur and Dufay, both consulted for the Bureau. Réaumur supplied expertise on the mining industry and championed the idea that the members of the Academy should serve as state administrators, while Dufay offered expertise on textile dyes.[29] The general controller of finances Jean-Baptiste de Machault d'Arnouville and Trudaine, his intendant, were both Academy's honorary members. There is no documentary evidence to determine whether Nollet gathered information on the silk industry in Piedmont during his 1739 visit. What is certain is that over the course of his first trip to Turin, Nollet came to believe that he would be asked to replace Dufay at the Bureau. On Nollet's mentor's death, however, the position went to the chemist Jean Hellot, to Nollet's great disappointment: "This position was promised to me before my return, and I was flattered with this expectation even more positively not more than three weeks ago. . . . I am not saddened, and I will not become, because of this, more active in running from waiting room to waiting room in order to seize fortune."[30] Although he did not work for the Bureau in any official role, Nollet served the state in a less public way. For several years, before and after his journey to Italy in 1749, Nollet acted a secret consultant for the Bureau in matters concerning the silk industry. The Bureau passed on to him reports on inventions concerning the production of silk threads, requesting that he deliver his opinion viva voce.[31]

Piedmont's Silk and the French State

In the eighteenth century, silk was a luxury commodity in high demand. Employed to make curtains and wall papers, to upholster chairs and sofas, to weave sheets and clothes, socks, and veils, it became even more valuable with the innovation of annual fashion changes. The French city of Lyon was Europe's silk capital.[32] Every year, Lyon designers created new fabric patterns that they proposed to Paris merchants, who then exported the goods to the rest of Europe. Once fashions took hold in Paris, a metropole that was synonymous with elegance and sophistication, the new fabrics quickly spread to other capitals. Nollet learned from a Lyon designer who shared a coach with him that the silk producers in Lyon feared counterfeits so much that they kept new models secret until the moment they reached Paris. The designer was on his way back to Lyon from such journey.[33]

The exports of silk fabrics made in Lyon constituted a flourishing business for the economy, yet there were aspects of the silk trade that troubled the French Bureau of Commerce. To produce their precious fabrics, the Lyon merchants did not rely on domestic silk threads but preferred those from the neighboring kingdom of Piedmont, in the northwestern part of Italy. In particular, they valued Piedmont's organzine—the thread that serves as the warp in silk weaving—above all others. The reason for their preference for a foreign product had to do with the perceived "superior quality" of Piedmont's organzine: the thread was uniform and resistant, and when woven, it produced better fabrics and less waste.[34] Travelers had arrived in Piedmont with the goal of stealing the secret of its successful silk industry since the first years of the eighteenth century. They came from France, Britain, and other countries, including neighboring Italian states. Britain even organized the migration of expert silk workers from Piedmont to its new colony of Georgia in the 1730s, and later to Bengal, in order to produce silk threads that would compete with those from Piedmont. In spite of large investment on the part of these imperial powers, however, Piedmont maintained its primacy in the silk industry throughout the century.[35]

The idea of "stealing secrets" underpins the romantic notion of the "industrial spy" as a cunning thief of technology. As Pamela Long and other scholars have demonstrated, however, the terms "secrets" and "secrecy" should be understood in their historical and semantic complexity, and not taken at face value.[36] Often, secrets were not secret at all. The *Dictionnaire de l'Académie françoise* published in 1777 explained that in the context of the arts and trades, for example, a secret was "a method known by a few people," such as blacksmiths, barber-surgeons, or other practitioners, "to make something, to produce some effects." In the realm of technology and industry, the word "secret" indicated "specific resources that can be put to various uses," such as, for example, the methods and processes involved in the manufacturing of silk threads.[37] These secrets could be donated, communicated, bought, or sold. Some of them were even published in "books of secrets," which were published compilations of alchemical recipes, medical remedies, and artisanal processes in vogue since the sixteenth century.[38] These "secrets" were clearly not secret.

Similarly, there was no real secret about Piedmont's silk. In 1724, the king of Piedmont issued printed regulations that determined each phase of the manufacture of silk. They included technical and legal details, such as the dimensions of each silk mill's and spinning machine's component, the salary system, and how and when apprentices should be trained.[39] By the late 1740s, travelers had brought to France copies of the regulations both in the original Italian and in translation along with models of mills and reeling machines.[40] Skilled workers too were enticed to France in spite of Piedmont's severe laws and harsh punishments to prevent spinners' migration. All the elements to recreate Piedmont's technology for the production of organzine were available in France well before Nollet's secret mission.

Setting up in France silk manufactures that would produce finished silk threads of the same quality as that of Piedmont, however, was not just a matter of stealing the secret of Piedmont's workshops. Even if the technical components and the skilled workers were available, to be competitive with Piedmont the French state needed to start large-scale production of high-quality raw silk, which was the real key of Piedmont's success. This upsizing of production was a political matter, not just a technical one. The French Bureau of Commerce needed to find ways of enforcing a system of production and quality control across its provinces, which entailed complex negotiations at the local level that had the potential to undermine the power of the central state.[41] These problems were absent in the much smaller kingdom of Piedmont.

At the beginning of the eighteenth century, François Jubié, the first in a dynasty of successful silk manufacturers in southern France, traveled to Piedmont "to study the different preparations of silk and to steal from Italy the secrets of its workshops." He enticed spinners to move to France, imported machinery, and established a manufacture whose silk threads reached the standard expected by the Lyon merchants.[42] Yet the machines and the spinners that Jubié imported from Piedmont did not solve the problem of establishing production at a large scale. In the 1740s, François's children petitioned the Bureau for state aid that would enable them to expand their activities to other provinces in a regime of semi monopoly.[43] The family's request had political implications that touched on the delicate issue of the relationship between the central state and its provinces.[44]

To tackle these problems, in 1741, the Bureau appointed the famous mechanician and inventor Jacques Vaucanson as the inaugural inspector of silk manufactures, whose role was to come up with a plan to reform the silk industry. Soon after his appointment, Vaucanson traveled to Piedmont to familiarize himself with the local silk industry and realized that its success hinged on strict enforcement of regulations that standardized production and ensured quality across its territory. Vaucanson did not advocate for the imposition of the same regulations in France; he proposed instead the creation of five royal manufactures, funded by a large company made up of the wealthiest Lyon merchants, who had an interest in locally produced, less expensive silk threads.[45] The royal manufactures would employ silk mills and spinning machines invented by Vaucanson himself, and would operate according to regulations that he prepared. Although the Bureau generously compensated Vaucanson's inventions, in 1744 this top-down "great design" backfired because of the violent riots organized by Lyon master workers, aware that the new regulations would cripple their business.[46]

The Bureau was in the delicate position of having to strike a balance among the various forces. The quality of Piedmont's organzine depended on the quality of raw silk, which was produced in large establishments where the state enforced a strict system of supervision and quality control. In France, instead, raw silk was produced mostly by *petits tirages*, small spinning workshops run by peasant families in the provinces, which were not easy to oversee or control.

Vaucanson's proposals, just like the Jubiés', aimed at eradicating the *petits tirages*, which produced low-quality silk and did not have any incentive to change their practices.[47] Intendants working in the French provinces, however, approached the matter differently. They sought to improve the quality of silk produced in the *petits tirages* by encouraging local inventors. The intendant of Languedoc Le Nain, in particular, strongly supported the invention of the abbé Soumille, a resident with "a talent for mechanics" who claimed that the introduction of a simple device he had contrived—a roulette—could produce high-quality silk without spinners having to learn new techniques.[48] Soumille's roulette could be employed in the small workshops as well as in large manufactures, and it was an economic alternative to Vaucanson's machines. Several manufactures had successfully adopted it in Languedoc, and Le Nain urged the Bureau to issue an ordinance that would impose it on all manufactures in his jurisdiction.[49]

Making decisions about which strategy to choose proved difficult for the Bureau because its members were aware of the political implications of their deliberations. To complicate things further, several reports that circulated at the Bureau suggested that France was missing important information about Piedmont's silk production and that the information the country did have was inaccurate. Crucial technical details, such as the exact number of teeth on the gears of Piedmont's spinning machines, were inconsistent in various reports from Italy.[50] It was in this climate of uncertainty that the Bureau turned to Nollet.

Nollet's mission was part of a larger scheme of state-sponsored intelligent travel. Since the days of Colbert the French state had considered manufactures and the arts as strategic domains to be kept under the control of the central administration. The Bureau appointed numerous inspectors and intendants to supervise manufactures in the provinces, with the aim to ensure fabricants' compliance with regulations.[51] Over the course of the eighteenth century, the political economy of the state became ever more dependent on the gathering of technical information through travel.[52] The massive 1715–18 regent's survey, consisting of thousands of reports on France's natural resources, arts, and manufacture, is a significant early example of the French state's investment in intelligent travel. Informants for the survey traveled to the French provinces as well as to various parts of the globe gathering information on the technical aspects of manufactures and on the organization of labor.[53] Decades later, Rouillé, a *commissaire du commerce* who later became the secretary of state for the navy, organized a systematic program of intelligent travel to foreign destinations. In his role, Rouillé supervised a group of young men, who received special education and became experts in the technical and commercial aspects of various manufactures. These trustworthy, well-trained civil servants offered technical expertise on a variety of commercial matters. Rouillé selected the most distinguished in this group to act as intelligent travelers. They visited foreign countries to gather information on various aspects of commerce or manufacturing, with the goal of helping France to import technologies that were absent or underdeveloped.[54] Nollet, Réaumur, and Dufay possibly traveled to Britain and the Netherlands under the auspices of this program.

Unlike Vaucanson, the Jubiés, or previous travelers, Nollet did not have any vested interest in the manufacture of silk. The Bureau believed that his report would differ from those that were already available, as it would offer impartial expert advice based on theoretical as well as practical knowledge. Nollet would provide instructions based on physical principles that, the members of the Bureau believed, would enable informed action on their part. The news of the electric cures performed by the Italians became a perfect cover for a secret mission the Bureau desperately needed. Nollet planned his departure so that he could be in Piedmont during the silk season. In advance of his departure, the Bureau forwarded to him a number of reports that explained the processes for the manufacture of silk.[55]

Secrecy, Dissimulation, and Science

While Vaucanson traveled to Piedmont in his official role as France's inspector of silk manufactures, only a small group of people was informed of the real reasons behind Nollet's journey. Everybody else was left to believe that he was traveling through Italy as a member of the Paris Academy of Sciences, interested in putting an end to an international debate on controversial electrical therapies. The Bureau instructed him never to write directly to Rouillé from Turin. He should send all letters to Rouillé's secretary so as to prevent any suspicion about the nature of his visit.[56]

This approach was similar to that employed in diplomatic espionage, yet Nollet did not proceed like a typical spy. He did not need to cipher messages, as Voltaire did in his letters to the minister of foreign affairs when he was in Prussia, nor did he find himself in the uncomfortable position of having to lie to the king of Piedmont, who warmly welcomed him back after ten years. The descriptions of his activities, which he shared with the Bureau, reveal a more sophisticated engagement with the culture of dissimulation typical of the early modern court. Nollet was especially careful to reassure Trudaine, the head of the Bureau, that he did not have to lie to Piedmont's king: "I have done nothing in Piedmont that might expose me to reproaches from the King, who honors me with his kindness. I haven't concealed any of my research or initiatives from H.M.; he knows everything from myself, and he has approved everything."[57] This apparently paradoxical declaration points to the intriguing ways in which the culture of disinterestedness that sustained academic and scientific reputations could be put in the service of secret affairs.

On his arrival in Turin, the king of Piedmont received Nollet and mentioned that he still remembered with pleasure the lectures and demonstrations of ten years earlier. As a sign of appreciation, Carlo Emanuele III offered Nollet an apartment at court, another in the city, and a servant. He also invited him to lecture again to the royal family and to join them on official business. Nollet alternated his duties for the Bureau with those as courtier in Turin. A few days after his arrival in Piedmont, he joined the royal family at the Venaria Reale, the royal hunting residence and a thriving site of silk production. He lectured, performed experiments, and waited after the royal family. In his spare moments, he gathered information on the early stages of silk production. At Venaria, he visited seven silkworm breeding sites, where he interviewed producers, commissioned drawings, and took extensive notes.

Back in Turin, Nollet resumed his lectures and similar commitments for the court, traveling to the various silk manufactures whenever he had a free day. When the king and the royal family went on a day trip, Nollet "took advantage of the time" to visit the countryside in the company of a military officer who was well acquainted with the silk industry and acted as an interpreter. Thanks to his help, Nollet was able to "ask a great number of questions" to local people so as to corroborate information he had gathered elsewhere.[58]

Nollet presented his visits to the silk manufactures as part of a more general interest in Piedmont's arts and trades. He asked to see the arsenal, the foundry, and the tobacco mills as well. He took notes on Piedmont's method of pulverizing marble, making caulks and bricks, rice and polenta, and various other productive activities. The king was aware of all his visits. He held Nollet's technical expertise in high esteem and invited him to evaluate newly designed fire extinguishers and several instruments destined for the royal cabinet.

Nollet dissimulated his interest in the local silk industry as learned curiosity. In his information-gathering activities, he employed norms of conduct that accorded with the ideal formula that Francis Bacon enunciated in his *On Simulation and Dissimulation*: "Openness in fame and opinion, secrecy in habit, dissimulation in seasonable use, and a power to feign, if there be no remedy."[59] Although it came under growing scrutiny in the eighteenth century, dissimulation was still widely practiced at Versailles. Books such as Nicolas Faret's 1631 *L'honeste homme* (*The Honest Man*), and Eustache de Refuge's 1649 *Traité de la cour* (Treatise on the court) explained that the art of dissimulation—the ability to disguise one's intention and to withhold information without lying—was essential in the life of the courtier. While lies were incompatible with gentlemanly status, dissimulation was an art that required self-control and presence of mind.[60]

Normative treatises on dissimulation published in the seventeenth century do not offer any insight on the actual practices of dissimulators. Nollet's travel diary and his letters from Italy to the Bureau, instead, offer firsthand accounts of the different registers that he employed while dissimulating. These documents reveal that he attended mass with the king on a Sunday morning and visited filatures in the afternoon, that he participated in the king's hand-kissing ceremony early one day and soon afterward met with informants to discuss silk matters, and that he squeezed electrical performances at court in between visits to silk workshops. These private sources demonstrate that Nollet's public reputation effectively disguised the real nature of his interest in silk. Nollet's status at the court of Turin, as well as his fame as a popular demonstrator and academician, allowed him to characterize his visits to silk manufactures as compliments he paid to his hosts' technical and economic success. He knew that as a respected savant with a commitment to improving the mechanical arts his interest in the local silk industry could easily pass as learned curiosity:

> I don't know what the people who had the complaisance to take me around will
> have thought of me: I saw them hundreds of times laughing up their sleeve over

the eagerness with which I walked into the workshops [*cassines*] to see in all stages those little creatures that work for men's luxury; these gentlemen will have undoubtedly thought that, by looking so many times at the same thing, I was curious to the point of childishness, and I think that they will have believed me strongly deprived of memory, since they saw me doing and repeating the same thing 20 times in different places.[61]

In the course of the two months he spent in Piedmont, Nollet dedicated twenty-three days to silk. He visited manufactures, compiled and commissioned detailed reports with drawings of reels and gears, measurements of components, analyses of the cocoon and mulberry-tree leaves trade, examples of common frauds, and descriptions of silkworms' maladies. He sent at least four folio illustrations of tools and machines employed in Piedmont to France, whose connection to those anonymously published in the *Encyclopédie* remains a matter of speculation as they do not survive in the archives. No one at court questioned the extent of his curiosity. The king of Piedmont even rewarded Nollet for his teaching and consulting activities with a knightly decoration, the cross of St. Maurice, which no previous Frenchman had ever received.[62]

Intelligence Networks

Like other travelers who reached foreign countries, Nollet relied on a network of translators, interpreters, intermediaries, and other kinds of go-betweens to access relevant information.[63] These intermediaries filtered the kind of knowledge that intelligent travelers obtained: they could misrepresent essential data or keep silent on crucial pieces of information. The ability to organize a trustworthy network of informants was a distinctive skill of a capable intelligent traveler. The constitution of such networks rested on an economy of exchange that was tacitly regulated by shared rules: a moral economy of secrecy.

The moral economy of secrecy required a diversified repertoire of rewards, which Nollet called "small gallantries," for each type of informant in the heterogeneous network he put together, which comprised academicians, merchants, owners of manufactures, and workers.[64] While bribing was effective with workers, monetary compensation would have insulted sources that inhabited the same academic world as Nollet. Self-identified citizens of the Republic of Letters shared ideals of free exchange and open circulation of knowledge that in principle set them morally apart from the world of trade where information was sold and bought. Disinterestedness and trust were primary values in the learned communities in which Nollet operated. Avoidance of pecuniary rewards characterized their transactions as cultural exchanges.[65] The abbé's main contact, for example, the recently retired physics professor at the University of Turin, Francesco Garro, joined Nollet's network in exchange for a position as a foreign correspondent for the Paris Academy of Sciences. Nollet, who had become acquainted with Garro during his first visit to Turin ten years earlier, knew that "in order to become a foreign correspondent for the Academy, which I have promised to him, he will do, and will do well, all I ask of him."[66]

Garro accompanied Nollet on his tour of the Italian peninsula acting as his interpreter, and was instrumental in recruiting Piedmont's royal mechanician, Isaac François Matthey, to Nollet's intelligence network. Matthey had obtained his position as a result of various strategic inventions in the military domain that remained state secrets. Prior to his appointment he had distinguished himself for his designs of machines and tools for silk manufacturers and merchants.[67] Although he held a royal appointment, Matthey received only a modest pension from the king and, at the time of Nollet's visit, he was preparing a formal complaint to demand more substantial rewards for his contributions to the state's prosperity.[68]

Rewards for inventions were regulated by patron-client relationships and were heavily dependent on the king's will and whims. Academic credentials could bolster inventors' reputations and enhance their chances of gaining the king's favors. Hence, with his promise to present Matthey's newly invented hygrometer to the Paris Academy of Sciences, Nollet knew that he would obtain the inventor's valuable collaboration. This was indeed the case: Matthey supplied Nollet with drawings of silk machines and measurements of each component, as well as details about the commerce of silk.[69] Matthey also provided an address in Geneva that Nollet could safely use for his correspondence with the Bureau.[70]

In their interactions with Nollet, Matthey and Garro trod a fine line between secrecy and openness. Both were public figures in Piedmont's cultural life. They contributed to the creation of an educational and research center based at the arsenal in Turin, which focused on the useful applications of experimental philosophy.[71] They offered Nollet access to technical knowledge as a peer, a learned savant with a strong interest in manufactures and the mechanical arts, without asking too many questions about what he was planning to do with the information he was gathering. It is unlikely that they did not realize what they were contributing to, but it was in the interest of preserving their own reputation to dissimulate in turn on the nature of their collaboration with Nollet. Visits to the arsenal or to state manufactures were not unusual requests on the part of learned foreigners. It was one of Matthey's duties as the royal mechanician to act as a sort of tourist guide for visitors interested in these aspects of Piedmont's life. Edward Gibbon was given such a tour in 1764 by Matthey and was impressed by his inventive ingenuity.[72]

In other contexts, however, Nollet bartered invention for information. His reputation as a skilled artisan made his insights on specific aspects of the silk industry valuable to Piedmont's producers. So, the owner of a silk manufacture introduced him to all the processes leading to the production of raw silk in exchange for the design of an improved furnace that would cut expenses for wood. In his correspondence with the Bureau, Nollet acknowledged "the innumerable curious things he kindly shared with me," labeling him an "honest man."[73]

If learned curiosity could effectively serve as a cover for the practical interests motivating intelligent travel, it did not work as efficiently in the world of trade and commerce, whose inhabitants did not share the same tacit values of the scholarly world. Accustomed to having to guard against frauds and cheating, Piedmont's merchants became suspicious of Nollet's insistent curiosity. After two months of visiting manufactures, Nollet reported to the Bureau that it had become "prudent to hide myself from the sights of the merchants in Turin, who began to gossip loudly about my research." He therefore traveled to Venice, where he resumed his role of the unbiased experimental philosopher fighting a battle against the love of the marvelous that plagued some Italian colleagues. He decided to cease communication with the Bureau, explaining to his superiors in Paris that merchants in Piedmont had mischaracterized his "innocent" inquiries: "A lot of people have gossiped, and if I engaged in correspondence with statesmen in France, they would not have failed to put my intentions in a bad light, thus making a crime of what I was innocently doing."[74]

The Value of Intelligent Travel

By insisting on the difference between his "innocent" curiosity and the suspicious world of trade, between the gossiping merchants and the welcoming gentlemen who openly showed him the processes of silk manufacture, Nollet wanted his superiors at the Bureau of Commerce to understand his report as the work of a savant, different from any other report they might have received.

He was no doubt hinting that he deserved to be rewarded more than the authors of these other reports typically were. In particular, Nollet hoped this kind of missions could result in his appointment as inspector or expert consultant within the Bureau. When he started planning his return to Paris, Nollet offered to complete his mission by visiting silk manufactures in the French provinces. He likely wanted to suggest to his superiors at the Bureau that he was ready to take on an official position in state administration or to consolidate his secret expertise.[75] The Bureau, however, again did not respond as Nollet hoped. While Nollet was in Italy, new policies as to how to best encourage French manufactures were developed after a reorganization in the top level of the Bureau. The arrival of Trudaine as the Bureau's director in 1749, followed by the appointment of Vincent de Gournay as intendant of finances, resulted in a distinctively more liberal approach to political economy.[76]

The new Bureau rejected the emphasis on regulations that characterized Rouillé's leadership in favor of an enlightened political economy of improvement aimed at educating manufacturers and merchants in the provinces.[77] This turn resulted in a more cautious consideration of Piedmont's heavily regulated system of production, which was the subject of Nollet's mission. The new Bureau's emphasis on improvement, education, and emulation had the effect of leading it to privilege Vaucanson's inventions and abandon the attempt to recreate Piedmont's system of production. In line with the more liberal orientations of the new Bureau, Vaucanson devised a plan that did not involve regulations but was based on machines that restructured the organization of labor. The plan was rewarded with the extraordinary amount of ten thousand livres.[78]

Vaucanson's new plan relied on updated information about Piedmont's silk industry, which he presented during a public meeting of the Academy of Sciences, just a few days before Nollet returned to Paris. Although there is no documentary evidence that Vaucanson used Nollet's reports to prepare this presentation, it is quite implausible that the Bureau would not have handed the new data from Piedmont to the inspector of silk manufactures. Vaucanson's new plan was a revised version of what he had presented after his travel to Piedmont in 1743 and demonstrated deeper knowledge of key processes for the manufacture of raw silk in Piedmont, which he certainly did not obtain firsthand. Even the timing of Vaucanson's presentation seems to indicate that the Bureau wanted to discourage the public from making any connection between Nollet's journey and Vaucanson's inventions.

Archival evidence indicates that Nollet was aware of how the Bureau used the reports he sent from Piedmont and that he resented its handling of his secret service. Once back in Paris, Nollet presented a detailed list to Trudaine of the "expensive steps I have been obliged to take in Piedmont and in Italy with respect to the research that Mr. General Controller and Mr. Rouillé entrusted me with." He underscored the merits of his mission, even if it now seemed less strategic to the Bureau: "If I have been charged with researching and collecting data that you already had, of which I had not been informed, I dare to say that I have as much merit as if I had discovered everything anew."[79] While in his letters from Italy he had emphasized the "new pleasures" that serving the state gave him, describing himself as the "happiest of physicists," confronted with the Bureau's delay in rewarding him for his secret service, Nollet emphasized instead his compliance with the Bureau's orders and the discomforts he had suffered while pursuing them.[80] He calculated that the Bureau should pay him about forty-five hundred livres, an amount that included expenses for transportation, lodgings, food, servants, and bribing informants within the silk workshops.[81] Compared with the ordinary per diem that inspectors received from the Bureau for their tours, the remarkably higher amount that Nollet requested indicates how special he believed his journey to be.[82] The Bureau's decision to round Nollet's compensation down to four thousand livres, however, less than the amount that Vaucanson had received in 1742 for a shorter journey, confirms that for the new administrators Nollet's reports were not particularly relevant.[83] Nollet's reports had been requested by a different Bureau, which believed in imposing regulations (hence in imitating Piedmont's system of production) more than in encouraging innovation and promoting emulation and improvement. The new Bureau invested in Vaucanson's inventions, which promised to produce high-quality silk threads with French technology.[84]

The new Bureau did not intend to waste the expertise the abbé had gained while in Italy. But it wanted to use it in secret. The reason for this secrecy was twofold. First, Nollet had traveled to Italy as a member of the Academy of Sciences who was interested in ending a controversy over the healing virtues of electricity. His reputation was strictly linked to that of the institution. The Academy itself would have been internationally censured had the secret reason behind his journey become public. Second, given the importance of silk and the numerous economic interests connected to it, it was advantageous for the Bureau to be able to resort to an expert whose identity was unknown to most. It was all the better that in the meantime, Nollet's reputation in the public arena escalated: in 1753 the king rewarded him with a new chair of experimental physics at the Collège de Navarre.

Academic Openness and State Secrecy

The expertise Nollet procured did become useful to the Bureau in the early 1750s, when the abbé Soumille submitted several proposals for improving the quality of domestic silk threads. Like Garro, Matthey, and Nollet himself, Soumille saw inventions as a means to enhancing his reputation and academic status. In 1737 he had been elected a corresponding member of the Paris Academy of Sciences in recognition for the several devices that he had sent to the institution for approval. In his quest for further appreciation, Soumille complied with the protocols of academic openness: none of his inventions was kept as a secret. On the contrary, working with the local intendant, Soumille organized public demonstrations comparing his roulette with Piedmont's spinning machine, styled on the theatrical experiments that contributed to Nollet's ascent in the learned world. The demonstrations took place at Le Nain's orangerie and confirmed that the two models worked equally well. As a result, Le Nain urged Trudaine to issue an ordinance that would compel all manufactures in Languedoc to use Soumille's roulette.[85]

Trudaine turned to Nollet in his assessment of the various options. Soumille had submitted a manuscript for publication in which he gave a mathematical explanation of France's faulty production of raw silk. Asked to determine whether the manuscript was worthy of publication, Nollet explained to Trudaine that Soumille's detailed descriptions constituted an open admission of France's inferiority to Piedmont and should be kept secret. However, he also suggested that the Bureau should take advantage of Soumille's expertise, recommending that Soumille "work in silence" for the state just as he himself had done.[86] Trudaine endorsed the recommendation. While Soumille had acted in accord with the rhetoric of openness that scientific academies throughout Europe had elaborated, the very same academicians who constructed that rhetoric knew from firsthand experience when to work in secret.

The activities of Nollet, Soumille, Matthey, and Garro indicate the persistence of an earlier culture of secrecy and dissimulation that coexisted with the culture of academic openness. Full-fledged citizens of the enlightened Republic of Letters mastered the arts of secrecy and dissimulation typical of earlier courts. Their public authority and expertise could be tied to less public—or indeed secret—forms of exchange and circulation. In the course of the eighteenth century, French administrators, who were often also members of the

Academy, became increasingly aware that academic status could conveniently lend an aura of gentlemanly disinterestedness to intelligent travelers. In the early 1760s, when Gabriel Jars was preparing for his journey to gather technical intelligence on the English mining industry, Trudaine made sure he obtained the official title of correspondent of the Academy before setting out on his tour. Jars regarded the position as "the most advantageous I could have when in a foreign country; they would suspect less an inquisitive member of an Academy than a man of the art."[87] Similarly, in the 1750s, the abbé de Sauvages, a member of several academies and an author on the natural history of silkworms, proposed himself to the Bureau for a tour aimed at gathering intelligence on silk manufacture under the guise of a "curious" person.[88]

All these requests, combined with Nollet's own secret mission, seem to indicate that the concept of industrial espionage was indeed emerging around that time. Nollet's insistence on the "innocence" of his travel, however, highlights the inadequacy of industrial espionage as an analytical category to account for the multifaceted culture of secrecy, dissimulation, and openness that characterized missions like his. Nollet did employ a number of methods and strategies used by diplomatic spies. He traveled under cover, bribed informants, and sent letters to the Bureau via mediators. On his return in Paris, he also made sure to dispel any lingering suspicion about the nature of his interest in Piedmont's silk by publishing a report on the cultivation of mulberry trees in Piedmont, based on his observations.[89] At the same time, the shared rhetoric on academic disinterestedness enabled him to dissimulate his interest in the silk industry as learned curiosity and "innocently" discuss it with colleagues and even with the royal family. Within the manufactures, however, several people were suspicious of such fabricated disinterestedness. Indeed, by the end of the century, learned entrepreneurs in Birmingham such as James Watt and Josiah Wedgwood had become quite wary of the "clever [French] scientific people" who wanted to visit their manufactures.[90] Their reluctance to open their workshops to the "curious" French indicates a new awareness that the imagined reality of academic disinterestedness could disguise more mercenary concerns.

Industrial Spies and Intelligent Travelers

During his visit to Bologna, Nollet noticed two effigies that hung from the tower in the main square. One represented Ugolino Menzani, a silk thrower who allegedly brought the hydraulic silk mill—invented in Bologna—to the nearby city of Modena in the late sixteenth century. Menzani was tried in absentia and sentenced to death as a "traitor of the motherland." The Bologna Senate regarded Menzani's activity as such a serious offense that it ordered a symbolic punishment in the form of the life-size puppet that Nollet saw. The symbolic punishment was reenacted in 1732, when another silk thrower who left Bologna for Venice was tried in absentia and sentenced to death. The second effigy that Nollet observed was his.[91] One may wonder whether Nollet felt any connection to these figures, who have been repeatedly described as "industrial spies." His brief annotation in his travel diary suggests not. He remarks on the "severity" of the Bologna Senate and refers to the two fugitive artisans as "those who take out of the motherland some of the arts that [the Bologna Senate] believe are the country's own."[92] This is in tune with Nollet's belief that technical knowledge was public good, a belief that shaped the course of experimental physics he offered to the Parisian elites and his participation in the encyclopedic projects of the Academy of Sciences. While the Bologna Senate regarded silk technology as its own property, Nollet saw knowledge as the province of all. He and his colleagues maintained that improvement—the perfecting of the arts—could be achieved by making best practices public and encouraging emulation among artisans and craftspeople. Nollet's remarks may also shed light on his characterization of his inquiries on Piedmont's silk industry as "innocent." His activities were not so much aimed at stealing secrets—as spies would do—but rather at gathering useful information as part of the larger scheme of perfecting the arts.

In the history of Italian silk various "industrial spies" take pride of place. Yet not even one among these imagined stories of industrial espionage has survived the careful analyses of silk historians. These episodes typically center on individuals who stole something from one place and brought it somewhere else. Historical accounts based on archival materials, however, show that technological transfers happened through collaboration and creative adaptations, usually over a long span of time. Hydraulic silk mills, for example, appeared outside of Bologna well before Menzani resettled in Modena.[93] The very beginning of Piedmont's successful industry is commonly attributed to an episode of industrial espionage that took place in 1664, when Gianfrancesco Galleani brought the hydraulic silk mill from Bologna to Piedmont. In this case too, scholars have demonstrated that the arrival of the hydraulic silk mill in Piedmont has a longer and more complicated history in which numerous social actors other than Galleani played a role. Piedmont acquired its undisputed superiority in the production of silk threads through small innovations in machine design and labor organization over the course of several decades.[94]

Yet the notion of industrial espionage and the related one of the industrial spy continue to overshadow the widespread practice of intelligent travel. Probably the most enduring story about silk and spies is that of John Lombe, an English mechanic who allegedly traveled to Piedmont at the turn of the eighteenth century "with a view of penetrating the secret" of silk throwing. In an account dating to 1791, his biographer related that Lombe remained in Piedmont some time and accomplished his mission "by corrupting the servants" and acquired all the knowledge he needed through "perseverance and bribery." Yet the plot was discovered, and so "he fled, with the utmost precipitation, on board a ship, at the hazard of his life, taking with him two natives, who had favoured his interest and his life, at the risk of their own." Lombe, the story continues, set up silk mills near Derby, what is widely celebrated as "the first modern factory." The revengeful Italians, however, sent an "artful woman," who seduced him and slowly murdered him by poison.[95]

This account of the Derby silk mills is largely fictional. Giuseppe Chicco has even questioned whether Lombe ever actually set foot in Piedmont, demonstrating that Piedmontese artisans who resided in London played a crucial role in the establishment of the Derby factory. The combination of espionage, seduction, and murder exerted a strong fascination on Victorian audiences. In the wake of the publication in 1821 of James Fenimore Cooper's *The Spy*, regarded as the first spy novel, numerous accounts of inventions and industry consolidated the imagined reality of the poisoned industrial spy that introduced one of the earliest mechanized factories into Britain. These accounts share a heroic approach to the history of science and technology, one that emphasizes materials, machines, individual inventors, and at times spies, as the driving forces behind industrialization and technological development. They also focus only on British industrialization, disregarding the case of Piedmont. Lombe's silk mills were—and still are—presented as "the first modern factory" in an egregious erasure of Piedmont's silk industry and a similarly egregious obliviousness of the failure of the British attempts to challenge Piedmont's primacy in the production of high-quality organzine.

By emphasizing Nollet's intelligent travel, my aim is to shift attention from the individual thief of trade secrets to the local dynamics that made the gathering of technical information possible. Nollet's mission was successful because he was able to rely on the collaboration of various Italian actors whose expertise he valued and trusted. The examination of such interactions, which are absent in the published accounts, casts light on the interest of imperial powers like France and Britain in the small state of Piedmont. When we put the accent on travel and information gathering, eighteenth-century Italy emerges as a place whose manufactures and learned communities made the crossing of the Alps a worthwhile effort. The absence of these interactions and activities from the printed page contributed to the propagation of imagined realities, which continue to shape narratives of early industrialization and the European Enlightenment, in which Italy was marginal unless framed within the paradigm of the Grand Tour.

CHAPTER TWO

Electricity, Enlightenment, and Deception

Electricity was one of the emblems of enlightened modernity. The "youngest daughter of the sciences," as the philosopher and theologian Joseph Priestley defined it, offered a vocabulary and a repertoire of embodied sensations with which to articulate visions of human progress.[1] The jolts and other involuntary spasms experienced by electrified audiences offered new metaphors for expressing the excitement of living in the era that Voltaire called the Iron Age.[2] Enlightened moderns turned the page on the nostalgic look to the past that characterized the previous centuries. Electricity was one of the tools they brandished to articulate a historical narrative that positioned them at the beginning of a new age of cumulative knowledge, material progress, and racial superiority. They conceded that the attractive properties that amber—"elektron" in Greek—acquired when rubbed had been known since antiquity but noted that only in the present had experimenters demonstrated that electricity was a universal power of nature. It was no chance that after "centuries of apathy" their age should witness a cornucopia of revelations about nature and its operations.[3]

One of the earliest British writers on the medical applications of electricity, Richard Lovett, regarded the phenomena of the Leyden jar, discovered in 1746, as nothing less than an act of divine revelation.[4] The Italian scholar Ludovico Muratori similarly declared that God "reserved for our times the discovery of a most wonderful phenomenon. I mean electricity."[5] Jean-Jacques Rousseau believed that if "a person had started up, in the last century, armed with all those miracles of electricity which are now common to the meanest of our experimentalists, it is certain he would have been burnt for a sorcerer, or followed as a prophet."[6] The German physician Johann Gottlob Krüger made the same point in his course of experimental philosophy dedicated to the crown prince, arguing that only a few centuries earlier electrical phenomena would have been considered magic.[7] One of the earliest proponents of the medical applications of electricity, Krüger also believed that the new science offered humankind powerful means to defeat diseases previously regarded as incurable.[8]

This staunch optimism in the breakthroughs that electricity would bring about continued throughout the century. When Benjamin Franklin championed the cause of the American Revolution, his contributions to the science of electricity became entangled with his political views. Lightning rods towered above palaces as banners of the human ability to tame natural and political excesses.[9] Even before the age of Franklin, the Republic of Letters had congratulated itself for living in an age of scientific progress, best exemplified by the new electrical discoveries. The community of the learned, the Italian scholar Francesco Algarotti wrote to Voltaire in 1746, was a cosmopolitan capital in which "eight to nine thousand people electrify one another."[10] Similarly, in his 1745 *Della forza attrattiva delle idee* (On the attractive force of ideas), the secretary of the Bologna Academy of Sciences, Francesco Maria Zanotti, described intellectual communities and scholarly disputes as caused, respectively, by the electric attraction and repulsion of ideas.[11]

The widespread success of the new science was facilitated by a preexisting experimental culture that crossed boundaries between academic and the social realms. Experimental demonstrations were part and parcel of the sociability of curious elites. Self-styled electrical performers staged spectacular shows of sparks and shocks for paying audiences, offering rational recreations and spreading interest in the new phenomena.[12] The "new fire that man produced from himself, and which did not descend from heaven" intrigued aristocrats and natural philosophers alike. Electrical science's success in the public arena is often associated to Franklin's theory and experiments, yet the early medical applications of electricity, which predated his work, contributed as much—if not more—to the popularity of the new science. Presented as "electricity made useful," medical electricity was an ever-present topic in publications dedicated to the science of shocks and sparks.[13]

Electrical experiments made news. Academic journals and popular magazines, along with a vast variety of other texts, popularized the surprising results of a heterogeneous group of self-styled "electricians" or "medical electricians," many of whom had just a smattering of natural philosophy and often no medical training at all. These publications excited readers' imagination, in some cases turning them into experimenters or healers. The Swiss physiologist Albert von Haller, remarked in a long article on electricity that was published in the *Gentleman's Magazine* in 1747 that electrical demonstrations had awakened the curiosity of those who would not normally pay attention to experimental philosophy. Not only the literate but even the "ladies and people of quality, who never regard natural philosophy but when it works miracles," became interested in electricity: "everyone wanted to see a "lady's finger darting flashes of lightning, or her charming lips setting houses on fire."[14] The vast number of publications on electricity, which employed a vocabulary emphasizing wonder and surprise, played a significant part in spreading interest in the new science.

Readers learned that electrified bodies acquired extraordinary properties. The practice of including the body in electrical performances dated from at least the 1730s, when Stephen Gray, a Fellow of the Royal Society of London, developed an experiment that demonstrated the ability of the human body to conduct electricity. His so-called flying-boy experiment became one of the popular demonstrations that made the fortune of itinerant performers in Europe and beyond (fig. 2.1).[15] In spite of the lack of agreement on the nature of the electric matter, electrical demonstrations involving the body became standard performances during fashionable electric soirées. They also offered evidence that electrification induced physiological responses, which paved the way for the medical applications of electricity.

This chapter reconstructs the context in which Pivati made his extraordinary claims about the medicated tubes and in which Nollet gained his authority as a leading expert on electricity. Its main chronological focus is on the few years (1744–47) during which the machine that made spectacular performances possible arrived in Italy from the German countries. I argue that while publications on electricity created an imagined reality in which electrical demonstrations could be easily replicated, it was in fact several itinerant demonstrators that made it possible for large numbers of Italian amateurs to perform experiments and advance theoretical claims about the role of shocks and sparks in the economy of nature. Academic interest in the new science grew in response to these traveling shows and promptly turned to the possibility that electricity might have medical applications. The common practice of experimenting on the human body made it plausible that electricity could be employed as a medical treatment, and in turn the possibility of medical applications boosted interest in the new science. The proliferation of self-styled electricians, however, created a crisis of authority within the experimental community, exacerbated by reports of prodigious electrical cures. The numerous electrical shows, the prodigious cures, and the flood of publications forged an unwanted connection between electricity and the charlatans and mountebanks that toured European cities trumpeting extraordinary medical remedies.[16] Electrical experts like Nollet strove to develop

N . le Sieur Invenit   R . Brunet fecit

Fig. 2.1. In this representation of the flying boy experiment from Jean-Antoine
Nollet's *Essai sur l'électricité des corps* (1746), the electrified boy, suspended from the
floor, uses his left hand to turn the pages of a book without touching it, while his
right hand attracts small pieces of paper. A lady makes a spark issue between the
boy's nose and her own finger. Note the aristocratic context of the scene and the
absence of the electrical machine. The machine is replaced by the electrified rod in
Nollet's hand, which looks like a magic wand. This is an idealized representation:
given the crowded space, an electrified rod would not have sufficed to produce all
the effects that we see in this image. Science History Institute, Philadelphia.

experimental protocols to create order among the overload of information—and misinformation—about electricity, its nature, and its effects.

Electric Bodies

All over Europe and its colonies, electrical performances shocked and instructed. They offered sensorial evidence for the existence of an all-permeating natural substance that did not reveal itself unless properly prodded. The spectacle of electricity relied on "electrical machines" that made the newly discovered power of nature tangible. Participants in electrical demonstrations experienced with their own senses the effects of the invisible power that performers initially called "electric fire." When connected to the electrical machine, audience members' bodies responded uncontrollably to the passage of electricity: their hands attracted small feathers or pieces of papers without their touching them, their hair stood straight up, and their entire bodies jolted when they underwent the "electric commotion" (fig. 2.2). Only following Benjamin Franklin's 1752 kite experiment did the association between

Fig. 2.2. Six people hold hands to receive an instantaneous discharge through a Leyden jar, while a female figure rubs the glass component of the electrical machine in this demonstration of the electric commotion in Peter Windler's *Tentamina de causa electricitatis* (1747). Notice the absence of the person (usually a servant) who would hand-crank the electrical machine. Collections of the Bakken Library and Museum, Minneapolis, MN.

lightning and electricity become widely accepted. Before then, there was no clear understanding of the nature of the electric matter.[17] The popular experiments seemed to demonstrate that electricity was an ever-present natural substance, even when invisible.

Electrical effects depended heavily on various circumstances, such as the humidity of the air, the quality of glass, the number of people in the audience, and the kind of clothing they were wearing. They were often difficult to replicate.[18] Published reports glossed over such difficulties, providing instead tantalizing descriptions of the effects of electricity on the human body, supported by an iconographic apparatus unlike that in any other experimental science. Images of electrified men, women, and children offered visual support to an imagined reality in which electrical demonstrations could be easily reproduced. The visual representations of electrified human bodies solidified in readers' minds the thrilling descriptions of electrical effects and generated shared expectations and understandings. In particular, this iconography erased the messiness of electrical experimentation: for example, the fatiguing operation of wheel cranking, usually relegated to servants, was rarely represented. In turn, these illustrations domesticized electricity as a natural, universal power and presented the unruly electrified bodies, depicted in the context of aristocratic salons and richly dressed, as socially acceptable.

While publications on electricity were sure to sell fast, electrical demonstrators played a crucial role in showing that electrical effects were natural phenomena. They were eager for their audiences to understand that the electrical machine did not create any effect, since it only revealed the electric matter that existed in nature. However spectacular, their performances were by no means to be confused with magic tricks. In *The History and Present State of Electricity* (1767), Priestley admitted that the phenomena of electrical attractions and repulsions "looked like the power of magic" and so, without explanation, "and with a little art," they could very well be used for "a deception of this kind." However, he underscored, the electrical machine was a "philosophical instrument," not a magician's tool, because it exhibited "the operations of nature, that is of the God of nature himself."[19] Just as the air pump—another philosophical instruments according to Priestley—demonstrated that the vacuum was a natural phenomenon, the electric machine manifested the electric fire, it did not create it.[20]

While experts debated the nature and properties of electricity without reaching consensus, electrified bodies seemed to provide evidence that electricity was an all-permeating, if still largely poorly understood, natural power.

The involvement of audience members' bodies was a distinctive feature of electrical demonstrations. In the various courses on experimental philosophy, audiences handled microscopes, orreries, and air pumps and, in line with Lockean pedagogy, used their senses to comprehend the laws of nature. Yet when the topic turned to electricity, participants' bodies themselves became part of the experimental set up. In his *Essai sur l'électricité des corps* (Essay on the electricity of bodies), published in 1746, Nollet offered the first comprehensive theoretical account of all known electrical phenomena, supporting his theory with a description of the physiological responses that any electrified audience member could feel. He believed that streams of electric matter—which shared some features with the substance of fire—issued in and out of electrified bodies. Participants in his lectures could use their own bodies to understand this theory: when they brought their cheeks close to an electrified object, they could feel the stream of electric fire causing a delicate tingling in their faces, and they could also see sparks issue from their fingers, hear cracking noises, and smell sulphuric odors.[21]

Electrified bodies jolted, whooped, and gasped, potentially disrupting the codes of civility and politesse of eighteenth-century sociability. Natural philosophers whose lectures attracted aristocrats, like Nollet in Paris and Georg Matthias Bose in Leipzig, devised strategies to make the uncontrollable electrified body socially acceptable. They turned electrical demonstrations into a new kind of group dance. Lecturers led participants in explosive choreographies that showed that the loss of control caused by electrification was only temporary. Just as men and women in the group dances performed during the social gatherings of elites were assigned different steps, so the most popular electrical demonstrations enacted well-established gender roles, which played with the elite culture of courtship and seduction, along with the sexual allusions connected to the vocabulary and gestures of electrical experiments.[22]

Bose, who performed for the duchess of Gotha, designed an experiment that turned ladies into electric Venuses. For this demonstration, which he called "Venus electrificata" and which Franklin later renamed the "electric kiss," a lady stood on an insulated stool while a gentleman tried to kiss her, only to receive painful sparks from her lips. Gentlemen, in their turn, could perform their virility by "inflaming spirits," that is, set fire to alcohol with electrified swords (fig.2.3). For those who preferred a less aggressive model of masculinity, Bose designed an electric "beatification," which made a luminous halo appear above a person's head (fig. 2.4). Nollet's celebrated "electric commotion," an experiment where people holding hands experienced an

Fig. 2.3. In this depiction from William Watson's *Expériences et observations pour servir à l'explication de la nature et des propriétés de l'électricité* (1748), the gentleman on the left is using an electrified sword to set fire to the alcoholic substance in the spoon held by a seemingly frightened lady. At the top, two children play with electrical attractions. On the right, another gentleman rubs the glass globe of the electrical machine, which seems effortlessly operated by a kneeling gentleman. Museo Galileo, Florence.

instantaneous electric shock at the same time, borrowed bodily gestures from the cotillion, a dance popular at court and among the aristocracy (fig. 2.5). The first and most spectacular iteration of such demonstration took place in the Hall of Mirrors at Versailles, the place for special balls, where Nollet invited one hundred eighty people to form a human chain.[23] Not surprisingly, Haller commented in his article that electricity had "replaced the quadrille" in the social gatherings of the time.[24]

Bose also introduced a gendered terminology into his theory of electric attractions and repulsions, which presented an innovative explanation of electrical phenomena based on the distinction between "male" and "female" electric fire. According to this theory, the male fire, emitted by metals and animal bodies, was strong and powerful, and sparks, with their crackling sound, were its visible manifestations. The female fire, on the other hand, was a weak luminous emanation, the kind of light that characterized the aurora borealis.[25] The gendered roles participants were assigned at electrical soirées in which

Fig. 2.4. In a darkened salon, a halo made of small sparks appears around the head of a seated gentleman, creating the electric beatification, in an illustration from Benjamin Rackstraw's *Miscellaneous Observations, Together with a Collection of Experiments on Electricity* (1748). The trick is the metallic device, in the shape of a crown here, which produces sparks and which is invisible in the dark. Collections of the Bakken Library and Museum, Minneapolis, MN.

Fig. 2.5. James Caldwall, *The Cotillion* (1771). Yale Center for British Art, Paul Mellon Collection, New Haven, CT.

entire families took part turned the potentially indecorous effects of electrification into socially acceptable choreographies.

Electrical demonstrations took place for the most part in the dark and capitalized on the gallantry and innuendo that characterized eighteenth-century sociability.[26] The gestures associated with electrical experiments—the rubbing of a long glass rod that emitted a stream of sparks or the gentle caressing of a globe—elicited the salacious curiosity of the salon goers and captivated the imagination of pornographers and satirists.[27] Theories that connected the electric matter to the principle of life, together with the discovery of animal electricity later in the century, further excited the popular imagination. The marquis de Sade was profoundly inspired by the violent bodily convulsions caused by the discharge of the Leyden jar and liberally employed electrical vocabulary in his works.[28] The idea that electricity could be used to promote fertility was at the core of the Temple of Health and Hymen, the London extravaganza of the medico-electrical quack George Graham. Among its many prodigious treatments, the temple featured a "celestial bed" surrounded by electrical effluvia, where couples allegedly could successfully conceive.[29] Authors who believed that the electric fire was connected to virility found short-lived confirmation in the rumor that the Leyden experiment did not work on the castrati.[30] Although the rumor was unfounded, the very fact that it spread reveals the pervasiveness of the sexualized interpretations of electricity and the related idea that "normal" bodies would respond similarly to electrification.

Italy Electrified

Eighteenth-century sociability provided the physical spaces and the social foundations for the science of electricity. In Italy, the electric craze started in 1746, during the War of the Austrian Succession, when a small group of traveling electricians from northern European countries, typically referred to as "Saxons," performed demonstrations in domestic and academic spaces. Equipped with the latest model of the electrical machine, still a rarity in the Italian states, these traveling demonstrators offered the opportunity to experience in real life the much talked-about electrical phenomena. They incited curiosity among publics that encompassed the academic and the amateur. Italian professors, members of academies, and dilettantes learned from them how to perform electrical experiments and then engaged in experimentation—for fun or in the hope of contributing new discoveries. The first electrical performer was Christian Xavier Wabst, a physician in service to the Austrian army, who arrived in Venice at the end of 1745. Wabst's performances attracted the interest of Giovanni Poleni, professor of experimental physics at the University of Padua, who wished to learn how to use the electrical machine. Since Wabst was bound to the Viennese court and could not leave Venice, he had to decline Poleni's invitation to teach him privately. Poleni was so impatient to learn that he sent his assistant to attend Wabst's lectures on his behalf.[31]

News of Wabst's activities quickly spread in learned circles in the various Italian states. *Dell'electricismo* (On electricity), the first Italian book on the subject, was published in Venice in 1746, and many people believed Wabst to be the anonymous author. "I'm curious to know who the main author of the new book *Dell'elettricismo* is," wrote the marquis Scipione Maffei to Poleni: "That Vapst [sic], the German doctor who was here, must have a part in it," he concluded.[32] In Bologna, too, there were rumblings that the work was by Wabst, even when the Venetian journal *Novelle della repubblica letteraria* (News from the Literary Republic) identified the Venetian physician Eusebio Sguario as the author of the two medical essays that concluded the text.[33] Giovanni Ludovico Bianconi, a Bolognese physician who was in Germany at the time the book was published, asked Maffei to send him a copy of "the work by Mr. Wapst."[34] Antonio Conti, a leading figure on the Italian intellectual scene in the early eighteenth century, was likely right in his belief that Sguario had composed the "book based on the reports of a certain Saxon who was well-versed in those experiments."[35] This statement is the most consistent with historical sources— Sguario claimed authorship in a later publication—and with the clues offered in the book itself.[36]

Another "Saxon" that toured several Italian cities was Francisco Bossaert, a Fleming serving in the Spanish army, who arrived in Venice in 1746, after attending Nollet's demonstrations at the Paris Academy of Sciences. Bossaert's spectacular performances impressed his audiences, who shared their excitement in their correspondence. Aristocrats, scholars, and learned women all contended for the attention of the "Flemish engineer." Five princes from Modena passing through Venice were so enthralled by his show that "the half hour lasted until two in the morning." The participants were eager to try out the effects of electricity on their own bodies: "The admirable spirit of the Prince of Este led him spontaneously to have himself electrified on the resins. . . . It was a marvel and particular pleasure to see a putto of ten years, unruffled, frank, laughing, allow himself to be touched on his cheeks, hands and legs and, when sending out sparks and scintillating, not complain or grimace at the touch and take painful offense at the fingers of others."[37] The young prince's composure changed the mind of the women who initially believed that electricity was harmful. During the same night, they "got involved, and by taking each other's hands they formed many semicircles in some gentle electrical experiments."[38]

The electrical choreographies were in part scripted, in part left to the improvisation of the performer. Bossaert brought along children assistants whose bodies he could electrify in case no one from the audience volunteered, whether because of fear or because they found it inconvenient. The display of the unruly electrified body had to comply with unwritten rules of proper conduct that varied from one group to the next. The difference in social status between audience and performer could result in awkward moments. On one occasion, "an experiment designed on the fly by the good Fleming" had to be suddenly interrupted. Bossaert wanted to electrify a five-year-old boy to show a phenomenon he demonstrated in other places, but as the secretary to the duke of Modena reported to Muratori, "as soon as I saw him kneeling near the child and after presenting him with a little earthenware basin place his other hand on his pants to unbutton them to have him urinate in public, I quickly drew close to him and whispered in French that he must refrain, because that experiment must not be done in front of girls."[39]

These kinds of shows promoted academic interest in electricity. *Dell'electricismo* gave plenty of directions on how to perform experiments, and it was so successful that it was issued in a second edition only one year after its initial publication. Yet textual descriptions, even with the aid of illustrations, were not sufficient to learn how to operate the electrical machine. Italian learned elites competed for Bossaert. Only after the Fleming taught him in his villa did Poleni finally learn how to independently perform electrical experiments. After Bossaert's departure in 1747, Poleni started to offer demonstrations for large audiences in the "theater of experimental philosophy" at the University of Padua.[40] Poleni performed for nobles, cardinals, officials, and tourists. Shortly thereafter, several Paduans bought electrical machines and dedicated themselves to electricity. Interest in the city was so widespread that Poleni had to discourage the Milanese demonstrator Girolamo Castelnuovo, who wanted to go to Padua to offer electrical demonstrations. Poleni explained that he was performing for the public at our university" and that "a gentleman from the Contuso family, a priest in the Danieletti family, and others have built several machines," and so he did not think there were enough people "curious to see electricity" who had not already seen it; such a trip, he concluded, "would be of no benefit to you."[41]

Poleni recommended Bossaert to his correspondent in Verona, the marquis Scipione Maffei, a linchpin of Italy's intellectual life.[42] Maffei was enthusiastic: he praised Bossaert as "a most honest man," "truly the gentleman" that Poleni had described him to be, worthy of the honor of sitting at his table.[43] Shortly after Bossaert had passed through, Verona sported eight electrical machines. Electricity "now kindles the most curiosity around here," admitted Maffei with satisfaction, recalling that in addition to himself there were "other *curiosi*" in town who amused themselves with it.[44] Maffei recommended Bossaert to his friend and correspondent in Brescia, Giammaria Mazzuchelli.[45] In Turin, Bossaert performed experiments for the king, queen, and royal children and had to decline the invitation of other nobles who insistently requested his services.[46]

Interest in electricity spread through the Italian states like wildfire. Traveling performers advertised their arrival in the Italian cities through preprinted publicity flyers that listed the most spectacular demonstrations. They filled in by hand the blank spaces with the details of the place, the duration, and the cost of their performances. One such flyer offers glimpses on the popularity of such shows and the business opportunities they provided. A Venetian woman called Lucieta Scanfarla hosted two three-hour electrical soirées daily for twelve days in her home; the entrance fee was roughly the same as that of a theater performance (fig. 2.6).[47] Wabst established himself in Venice at the end of 1745, and by the beginning of 1747 Bossaert had traveled all over northern Italy, while other "Saxons" had traveled through Tuscany and the southern part of the Italian peninsula, literally sparking interest in electricity anywhere they visited. They reached capitals and provinces. In Perugia, a city that hardly ever appeared on typical Grand Tour itineraries, the electrical machine arrived in 1746. Prospero Mariotti, a physician who taught theoretical medicine at the local university and offered lectures on experimental philosophy to aristocratic women, promptly enriched his repertoire of demonstrations with sparks, attractions, and electric commotions.[48]

In Rome, it was only after the demonstrations of a "Saxon professor" that the electricity craze took hold. A modern electrical machine had arrived in the city from Leipzig in 1744, but not much experimental activity had ensued. The "Saxon," however, elicited the "the admiration of all classes," including cardinals and university professors, who subsequently dedicated themselves to electricity.[49] At the Roman College, the electric shows impressed Benedict Stay, the author of didactic poems on Cartesian and Newtonian philosophy, who included electricity in the 1747 edition of his lengthy didactic poem entitled *Philosophiae versibus traditae libri vi* (Six books of philosophy delivered in verse). He described the various experiments performed by an "expert . . . who tours cities and kingdoms" in order to instill the love of truth in the ignorant, discredit alleged miracles, and expose superstitions.[50] Stay was one of several Roman Jesuits who became interested in electricity after the "Saxon professor" performed at the university. Giuseppe Bozzoli, another professor at the Roman College, started to perform electrical demonstrations for the public soon after the Saxon left Rome, regularly attracting crowds of spectators, both aristocratic and otherwise.[51] The same Saxon performed in Bologna, Piacenza, and Modena, where he enthralled Muratori, among others.[52] In Naples, interest in electricity erupted in 1747, when the "Saxon" Johann Peter Windler performed in the library of the prince of Tarsia, Ferdinando Vincenzo Spinelli, in the presence of the Neapolitan elite. Windler's electrical machine caused a stir: "Everybody flocked around it; everybody admired it as if it were a miracle."[53]

Because of its novelty, electrical science did not require extensive training or education and allowed a vast range of dilettantes to pass as experts. Electrical

NOBILISSIMI SIGNORI.

R Eſtano invitati, e pregati a volerſi degnare d'intervenire

R St Carla en cura Dela Scia Lucietta Scanfarla

Dove ſi fanno vedere molte Eſperienze di Elettricità non mai più vedute, come per eſempio:

    1. Di fare ſortire del Fuoco dal doſſo di una Perſona.

    2. Di far comunicare detto Fuoco da una Perſona in un' altra.

    3. Di fare arricciare i Capeli di qualche Perſona.

    4. Di far ſortire del Fuoco dall' Acqua, reſtando la medeſima Acqua fredda.

    5. Di mettere qualcheduno in ſtato di accendere lo Spirito di Vino con un dito, ſenza toccarvi.

    6. Di far condurre un piccolo Vaſcello ſopra l' Acqua, con un dito, e ſenza toccarlo.

    E finalmente molte altre Eſperienze aſſai rare, e ſingolare, e che hanno reſa ammirazione ben grande a tutti i Perſonaggi, e Virtuoſi di Europa, che le hanno vedute.

⧫⧫⧫⧫⧫⧫⧫⧫⧫⧫

Il Prezzo, e Pagamento ordinario è di una lira pr testa

Le Perſone Nobili, e qualificate daranno quel tanto, che parerà, e piacerà alla loro Generoſità.

Si darà principio il dì 18 *del corrente Meſe di* 9br 1746 *fino al dì* 30 9bre *la mattina dalle ore* 16. *ſino alle ore* 19., *ed il dopo pranſo dalle ore* 22. *ſino all' un' ora di notte.*

Fig. 2.6. This itinerant electrician's flyer reads "Noblest gentlemen, you are invited to, and please do join, the house of Ms. Lucietta Scanfarla in St. Clara, where you'll be able to see many new experiments of electricity, such as: 1. The extraction of fire from a person's body. 2. The communication of this fire from one person to another. 3. A person's hair being ruffled. 4. The drawing of fire from water, even as the water remains cold. 5. Wine being set on fire without its being touched. 6. A little ship being led over water with a finger without the finger touching it. And, finally, many

instruments were often built in-house by local artisans, who, upon receiving guidance from the foreign demonstrators or their pupils, repurposed existing spinning wheels or lathes to satisfy their clients' new interest. Cello strings and glass vessels from Venice completed the experimental setup (fig. 2.7).[54] In the aftermath of Bossaert's departure from Verona, Maffei had an electrical machine built and told his London correspondent Richard Mead that "there would be no end if I told you how much work has been done in my home on this account, and how many experiments have been attempted."[55]

Maffei shared his enthusiasm for things electrical with the count Gazola and his wife, the countess Massimiliana Guarienti. Having "taken a fancy to electrical experiments," they too had a machine built (fig. 2.8). The count, who was known in Verona for his invention of a device to protect the banks of the River Adige, had set up a room for experiments in his villa, which became a research center and a venue for learned sociability. Unpublished drawings document the wide range of electrical experiments performed in the Gazola-Guarienti laboratory. They show that electrical soirées were part of the social life of this group of aristocratic men and women. They performed experiments and engaged in conversation when they took a break. Research and sociability went hand in hand in their villa. The drawings may have been mnemonical tools used to set up the demonstrations or, more likely, preparatory steps toward a publication that never materialized. Gazola and Guarienti hoped to measure the speed of the so-called electric fire and to find a quantitative relationship between the number of rotations of the machine's wheel and the flow of electric fire emitted from the glass. While seeking to discover a new law of nature, participants in the Gazola-Guarienti laboratory also entertained themselves with electric commotions and exchanged electric kisses (figs. 2.8–2.11).[56]

Fig. 2.6. (continued)

other unusual and rare experiments, which have sparked great admiration in all the greatest and the most virtuous in Europe, who have seen them. We will start on the 18th of the current month of November 1746 [and continue] until the 30th of November at 16 in the morning until 19 and after lunch from 22 until 1 at night. The price and ordinary payment is one lira per seat. Noble people, and those of quality, will give what will please their generosity." The cost of the lecture was roughly the average amount a salaried worker received for a day and a half of work, but it was much less than the price of an average book. The time was measured in Italian hours. It corresponded, roughly, to 9–12 a.m. and 3–6 p.m. in current time. Biblioteca nazionale Marciana, Venice.

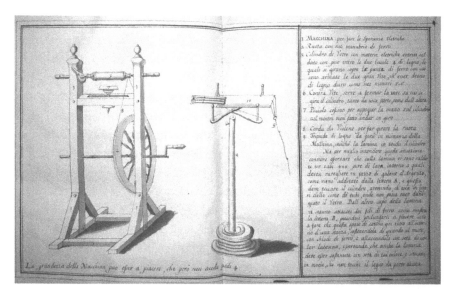

Fig. 2.7. A drawing of marquis Luigi Sale's electrical machine with instructions on how to build it. Bibliothèque municipale de Nîmes, Nîmes. The list to the right mentions a "violone string to make the wheel turn."

The domestic context in which experiments took place facilitated the participation of women in electrical science. While we do not know whether Lucieta Scanfarla was personally interested in the new science or simply saw a profitable opportunity, Mariotti's *Lettera scritta a una dama sopra la cagione de' fenomeni della macchina elettrica* (Letter to a lady on the cause of the phenomena of the electrical machine) shows that his female students actively engaged in electrical learning, questioning him and challenging his arguments. Electricity was certainly a powerful propeller for the more academic trajectories of Laura Bassi in Bologna and Mariangela Ardinghelli in Naples. Ardinghelli came to the attention of the local community when Windler performed electrical experiments at the presence of Naples' "brightest lights." She was the only one who "had the courage" to comment and ask questions. Subsequently, she became one of the Italian learned women whose international reputation attracted the curiosity of foreign travelers.[57] Bassi, who was the first woman to hold a university professorship, was at the center of electrical experimentation in Bologna along with her husband, Giuseppe Veratti.[58] After their marriage in 1739, Bassi and Veratti opened an academy dedicated to scientific topics in their home. The academy met twice a week and attracted numerous "foreigners, nobles as well as men of letters," who learned from Bassi's experimental demonstrations.[59]

Fig. 2.8. In this rendering of the Gazola-Guarienti electrical machine, the gentleman seems to be turning the handle while also rubbing the glass element. Biblioteca nazionale Marciana, Venice.

As in other European countries, in Italy too electrical vocabulary and imagery came to be associated with love and seduction. Bassi and Veratti were both members of the Accademia dei Vari (Academy of the Varied), which met in the palace of the senator Filippo Carlo Ghisilieri and became the arena for Zanotti's electric poetry. In 1746 Zanotti published *Amore filosofo* (Love, a

Figs. 2.9–2.11. These details of electrical experiments in the Gazola-Guarienti laboratory come from a larger poster that represents several other experiments. At the top, the ruffling of hair demonstration takes place as the lady, presumably the countess Guarienti, is engaged in conversation. In the middle, the inflaming of spirits experiment is performed on the countess herself, while two noblemen are engaged in conversation. At the bottom is the electric commotion. Notice the similarity between these sketches and published images of similar experiments. Biblioteca civica di Verona, Verona.

Fig. 2.12. In the frontispiece to *Amore filosofo*, a couple of putti are effortlessly operating an electrical machine that resembles a music box. Notice the bow and arrows at the bottom left. Biblioteca comunale dell'Archiginnasio, Bologna.

philosopher), a collection of three poems that celebrated the wedding of one of the members of the Academy, in which he described how Cupid, the young god of love, on discovering the electrical machine, hung up his bow and arrows and instead electrified the lovers' hearts.[60] For the moderns, the spark of love was an electric spark (fig.2.12).[61]

Medical Electricity

Correspondence between the men and women who were involved in experimental physics—as amateurs or professionals—reveals that they frequently exchanged information about the electrical demonstrations that took place in universities, monasteries, courts, villas, and palaces. Beyond the spectacles of sparks and shocks, what really inflamed their minds was the possibility that electricity might open up a new branch of medicine: medical electricity. This possibility emerged from the very domain of spectacular demonstrations, which routinely indicated that electricity induced physiological responses. The extraction of sparks from arms, fingers, or other body parts provoked painful sensations and involuntary movements that elicited questions as to the effects of electrification on the human body.

In the 1730s, the French naturalist and Nollet's mentor Charles de Chisternay Dufay noted that his arm went numb as a result of the numerous sparks he extracted when he was experimenting with electricity. In the following decade, reports of this kind multiplied. Sguario and Wabst communicated that a man with a "sad soul" and a "lazy and heavy body" felt cheerful and lighthearted after participating in electrical demonstrations.[62] Leipzig professor Christian Hausen observed that multiple small red dots, similar to flea bites, appeared on the body of a young girl, who—"joyful and fearless"—had undergone electrification in the course of an electrical soirée. Similarly, Krüger stated that all the people who subjected themselves repeatedly to the "electric kiss" had small red spots on their hands.[63]

Krüger had somewhat ironically stated that "since electricity must have some utility and this cannot be sought in theology or law, all that remains, obviously, is medicine."[64] Indeed, the possibility that the newly discovered power of nature could have useful applications did boost public and academic interest in the new science. Several experimenters, some of them physicians, investigated the nature of the pain induced by shocks and sparks and other physiological effects. Together with his pupil Christian Gottlieb Kratzenstein, Krüger was one of the earliest proponents of the medical applications of electricity. He noted that electrification produced involuntary muscular motion and suggested that it could be employed to treat palsies.[65]

Kratzenstein applied electric treatments to several patients and found that their pulse increased by a third following electrification.[66] Bose too conducted medical experiments on electrified patients and concluded that electrification increased blood circulation, insensible perspiration, and pulse rate. He claimed that all diseases that benefited from such effects could be treated by electricity.[67] In support of his views, Bose designed a simple demonstration, called the "electric siphon," which showed that a quantity of water that initially dropped out of a siphon flashed out in full stream upon electrification (see fig. 2.14). This, he maintained, explained the accelerated circulation of bodily fluids under the effect of electricity.

In these early phases, the administration of electricity as a treatment did not require specific instrumentations or techniques. Sguario and Wabst performed bloodletting on a man suspended as in the flying boy demonstration to test the effects of electrification on the circulation of the blood.[68] Lovett mentioned three ways for using electricity as a medical treatment, none of which were particularly different from the ways electricity had been used in various demonstrations involving the human body. The main difference was that medical applications took longer and were repeated over the course of several days or even weeks.[69] This meant that virtually anyone with a basic set of electrical instruments could become a "medical electrician." Presented as successful when everything else failed, medical electricity acquired the traits of a God-sent cure in certain publications. In England, religious groups embraced medical electricity as evidence of divine benevolence and as a second revelation.[70] One of the earliest supporters of electrical treatments, John Wesley, presented electricity as an original remedy created by God before the Fall, only recently revealed to humankind.[71]

In the Italian states, several of those who trained themselves by hiring traveling electrical demonstrators turned to medical electricity as soon as they were able to perform independently. In Padua, the celebrated anatomist Giambattista Morgagni together with Poleni investigated whether electricity could have useful applications in pyrotechnics or in "what is most hoped for, medicine." Poleni started an experimental program focused on the effects of electricity on the human body shortly after Bossaert's departure. He measured the pulse during and after electrification on several individuals, including himself. His results confirmed the common notion that electricity accelerated the pulse rate. He concluded that electrification could benefit "all illnesses arising from turbid blood" and decided to test this result on a paralytic man, who showed slight improvements after being electrified for about half an hour twice a day for four days.[72]

One of Poleni's correspondents, the marquis Luigi Sale of Vicenza, also dedicated himself to electrical treatments. On Poleni's recommendation, Sale invited Bossaert to visit him, and shortly after the visit he had an electrical machine built and started self-guided experiments (see fig. 2.7). Poleni's results with respect to pulse rate prompted Sale to apply electricity to a servant of his, who had become a local celebrity because of his remarkable sleepwalking. Better known as the "sleepwalker from Vicenza," Sale's servant Giambattista Negretti got out of bed every spring night and, without waking up, tried to eat and drink, borrow money, go to the tavern and, "as is often the custom of these sorts of people, especially in our city which abounds with choice wines, shamelessly drink." Several physicians had written about this particular case. They had resorted to various, sometimes harmful, treatments that, while entertaining the marquis and his friends, invariably failed to wake up Negretti. Having learned about the medical powers of electricity, Sale asked the physicians to extract sparks from the sleepwalker's body. The treatment brought about the desired effects after only a few attempts. The physicians reported that the sleepwalker recovered almost instantaneously and was so happy that he not only thanked the marquis "with tears of consolation" but also generously blessed the electrical machine.[73]

The medicated tubes, Pivati's controversial invention, emerged from this climate of enthusiasm for electrical phenomena and optimism about their useful applications. As the next chapter details, Pivati too attended Wabst's and Bossaert's lectures in Venice and soon turned to medical electricity. As a field of practice that joined medicine and physics, however, medical electricity occupied a liminal space that affected its reception. Some physicians regarded electrical experimentation as akin to the mechanical arts and thus as having little to offer to medicine, a liberal art, while other medical professionals were optimistic about the new science and joined experimental physicists in the attempt to assess the therapeutic virtues of the newly discovered electric matter.

In 1746, Nollet, along with the physician Jean-François Morand and the surgeon Joseph Marie François de la Sone, started a series of systematic tests to determine whether electricity could be of use in the treatment of palsies. The group applied electric shocks to paralyzed patients, but while the patients felt some pain and even a little bit of itching in the electrified parts, their responses appeared temporary and too varied for any definitive conclusion.[74] In Geneva, the professor of experimental physics Jean Jallabert applied electricity to a blacksmith named Nogues, under the supervision of members of the local medicine and surgery faculty. Nogues, who had been paralyzed for fourteen years, underwent electrification for an hour and a half every day, for about forty-five days, and was successfully cured. To emphasize that the result was more than a temporary effect, Jallabert supplied Nogues's address in Geneva and explained that anyone could visit him and see the positive effects of electrification for themselves.[75]

Accounts of Nogues's cure circulated widely in print, supporting the notion that electricity could be used as a medical treatment. Various published reports detailing seemingly prodigious electrical cures prompted a variety of individuals to administer electricity as a medical treatment, both under and—above all—outside the supervision of the medical profession. The financial opportunities that medical electricity offered were not lost on academic electricians, who often put questions regarding therapeutic efficacy to one side to pursue lucrative activities. Boissier de Sauvages, royal professor of medicine at the University of Montpellier, devoted himself to studying the therapeutic effects of electricity soon after reading about Nogues. He reported to Jallabert that the cure of the paralytic from Geneva had made electricity fashionable in Montpellier. Several individuals had portable electrical machines built and offered treatments. "In this town everybody gets electrified," he concluded.[76] Nollet too corresponded with Jallabert about the growing interest in anything electrical. He emphasized that the news of Nogues's cure made electricity "sell really quickly."[77]

Testing Medical Electricity

The proliferation of self-styled medical electricians and the stories of extraordinary electrical cures published in dedicated booklets or magazine articles made it difficult for the academic community to take a firm position on the efficacy of medical electricity. In 1744, the physician Cromwell Mortimer, who was the secretary of the Royal Society of London, shared his hope that future experiments would reveal that "electricity may be used medically."[78] The naturalist Henry Baker presented a lengthy report to the society on the various cures that numerous individuals all over Europe claimed to have performed, including Pivati's, concluding that the medical applications of electricity "romantic as it might seem, should not be absolutely condemned without a fair tryal."[79] William Watson, however, the society's most authoritative electrician, did not have the same confidence in the practical applications of the new science. He was by no means an exception. Physicians like Sguario and Wabst regarded medical electricity as an intrusion into their domain, and in *Dell'electricismo*, they adopted a critical stance regarding the presumed curative virtues of electricity. Even Luigi Galvani, the celebrated Bologna physician who experimented on animal electricity in the 1780s, seems to have never administered electricity to his patients.[80]

To complicate matters further, in the eyes of many, medical electricity had the air of charlatanism about it. Self-styled electrical healers, many of whom toured various cities, advertised their treatments as a miraculous last resort remedy, targeting patients who had already undergone official therapies without experiencing relief. The similarity between electrical demonstrations and charlatans' traveling shows was clear to several observers. The Venetian *Novelle della repubblica letteraria* criticized the "practical experimenters, or itinerant philosophers, who, sponging off the electrical virtue, have found a livelihood by wandering around the world with some electrical machine they have crudely slapped together, as we have seen done all over Italy."[81]

The Jesuit Jacopo Belgrado echoed the complaint about the lack of expertise of many self-styled performers, pointing out that "electrical phenomena have become so commonplace and popular nowadays that even the lowliest, most uncouth of common folks boasts of having observed them and feels entitled to throw in his two cents' worth, as if pontificating with the keenest of philosophers."[82] The physician Giovanni Bianchi extended the criticism to the learned, lamenting that the "electric virtue now clutters the minds of almost all philosophers, so that among themselves they speak of nothing else."[83]

This loud impatience toward the widespread excitement for medical electricity casts light on Sguario's decision to publish *Dell'electricismo* anonymously. Embracing publicly a nascent science that was still in the hands of traveling demonstrators would have been risky for a physician like him. In Naples too, an established experimenter like Giovanni Maria Della Torre preferred to err on the side of caution when it came to authorship. Soon after Windler's departure, Della Torre collaborated with the printer Porsile on the publication of a new book on electricity. Capitalizing on the electric mania that Windler's arrival had stirred in Naples, they assigned authorship to Windler yet sent a more ambiguous message in the accompanying plates. Unlike similar illustrations in other works on experimental philosophy, the plates in Windler's text were produced with an attention to detail that made both the place and the people identifiable to the local audience. The main plate, in particular, represented Della Torre in the role of Windler in the palazzo Tarsia, as he operated the electrical machine with Ardinghelli and others. For the local audience, who could easily identify the portrayed people, this plate indicated that Della Torre was the book's author, yet nowhere else was Della Torre's role in producing the book explicit. In the following years, Della Torre privately claimed authorship of the book, while in public he more cautiously stated that he merely added annotations to Windler's text (fig. 2.13).[84]

For experimenters like Nollet, who regarded electricity as a new career opportunity, academic skepticism about the status of electrical science was a serious challenge. Nollet was wary of the various "schools of electricity" that had emerged in Europe. They circulated conflicting reports on electrical experiments that undermined the new science in the eyes of the academic community. "Nowadays, everyone busies themselves with electricity," he remarked, adding that it would make sense to ban such studies or limit the freedom to write on electricity. Since enforcing such limitations would be impossible, though, Nollet decided that the only recourse against deception and misinformation was a systematic experimental program aimed at

Fig. 2.13. Electrical experiments at Palazzo Tarsia in Naples depicted in an illustration in Peter Windler's *Tentamina de causa electricitatis* (1747). The man that rubs the glass component of the electrical machine looks very much like Della Torre as rendered in contemporary portraits of him. The woman can confidently be identified as Mariangela Ardinghelli, the only woman who regularly participated in the experimental activity at the palace. Notice that the electrical machine is operated by a servant. Collections of the Bakken Library and Museum, Minneapolis, MN.

establishing the effects of electrification on organized bodies once and for all. The possibility that electric shocks could restore movement to paralyzed limbs was an idea that "comes naturally to mind," he declared, but any conclusion on efficacy required the collaboration of experimental physicists with medical professionals.[85]

Nollet approached the conflicting hypotheses on the effects of electrification by marginalizing the role of the patients' subjective reports and focusing instead on evidence offered by instruments and measurements. He based his work on Kratzenstein's conclusion that electrification increased perspiration and accelerated the pulse and on Bose's experiment on the electric siphon. By means of a specially built scale, he weighed plants and animals before and after electrification, noting any weight change. He then repeated the same measurements on human bodies. All his tests showed variations in weight after electrification. Human bodies, in particular, underwent a measurable loss of weight due to increased perspiration. Nollet presented these results as a starting point for further investigations that would have to be completed by physicians: "It is up to the medical arts to examine and test if this new way of increasing or causing perspiration and flushing the pores of the skin is equally beneficial to the sick, as it poses few if any discomforts to the healthy: because it is quite certain that neither I nor those who have assisted me have ever experienced any other discomfort, apart from feeling a little low and a big appetite."[86]

Nollet confirmed Bose's result that electrification caused an acceleration of fluids moving in the capillary tubes and suggested that this effect could be employed to remove obstructions. Yet he reiterated that it was up to "those who possess the art of medicating . . . to decide under which circumstances, without the slightest fear, one might force the perspiration needed for diseases."[87] His caution underscores the extent to which experimental philosophers who were not themselves physicians were careful not to claim medical expertise while experimenting with electricity. Medicine was a prestigious liberal art and interfering with it without explicit invitation could disqualify even an expert experimenter like Nollet as a charlatan.[88]

Nollet was busy with this experimental program when he heard the news of Pivati's prodigious healing of the bishop of Sebenico and of his experiments that seemed to demonstrate that odorous substances trapped inside a sealed glass tube could be diffused in the surrounding air when the tube was electrified. Pivati's results were a perfect match for the kind of investigation Nollet was already undertaking. In principle, Nollet did not exclude the possibility that the electrified substances could pass through glass. According to his conception of matter, every material body consisted of solid parts between which there existed empty spaces called "pores."[89] This was also true for glass. Electricity, together with heat and light, was one of the "subtle fluids"—that is, not composed of ordinary matter—that were "imponderable," meaning that they had no weight and so could pass through these pores.[90]

Nollet's own theory of electricity was based on the possibility that electric matter could cross freely through glass. However, Pivati's results seemed to indicate that substances placed inside the tubes (made of ordinary matter) could also pass through the pores of the glass. This was a claim that Nollet's scale could verify or prove false. Nollet reported that after rubbing the tubes filled with Peruvian balsam "thirty times" and on different occasions, they never gave off any odor and the scale detected no weight loss. Nollet concluded that no ordinary substance placed inside a sealed glass tube came out of it as a result of electrification. By the end of 1747, two years before his Italian journey, this was no longer a controversial issue for him (fig. 2.14).[91]

Electric Deceptions

Popular magazines, booklets, and other publications promptly spread news of prodigious cures and self-cures effected by electricity. In 1747, the *Gentleman's Magazine* reported that during an electrical soirée a man tested the effects of electricity on his rheumatic finger, finding great relief. Accounts like this quickly proliferated, along with publications advertising electrical treatments.[92] On the printed page, electricity became a new panacea, able to cure any kind of disease, from fevers to tumors, through rheumatism, gout, palsies, nervous disorders, constipation, "obstruction of the menses," along with the problems that typically fell under the care of medical charlatans: those pertaining to eyes, ears, and teeth. Published accounts of electric cures typically described the unsuccessful previous treatments, together with the duration and type of electrical application. In some cases, they gave detailed descriptions of sensational cures—especially prodigious because the disease itself was either deemed incurable or remarkable.

On the printed page, the sleepwalker from Vicenza, together with the paralytic from Geneva and the bishop of Sebenico, became celebrated testimonials—yet by no means the only ones—of the healing virtues of electricity. Where the cures happened—the house of a marquis, say—or how they were framed—attested to by a physician or a bishop, for example—enhanced the credibility of the narration. For most readers the electric cures became matters of fact: the identity of the patient, of the healer, the witness, or the reporter lent credibility to the reports.[93]

In the early modern period (which I take to include the eighteenth century), voicing skepticism about an experimental result was equivalent to challenging the experimenters' and the witnesses' authority. It was a bold move. It is no coincidence that those who had strong doubts about the case of the

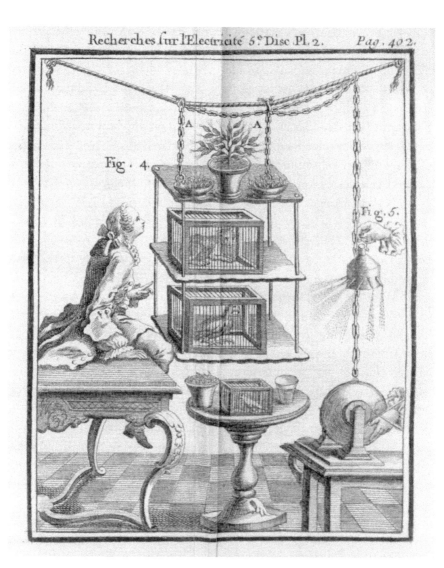

Fig. 2.14. Jean-Antoine Nollet's experiments on "organized bodies" as depicted in his *Recherches sur les causes particulières des phénomènes électriques* (1754). Note the electric siphon held by the hand on the right, and the elegant setting in which the experiments take place. While disembodied hands operate the electrical machine, the observer wears rich clothes. Medical Historical Library, Harvey Cushing/John Hay Whitney Medical Library, Yale University, New Haven, CT.

sleepwalker from Vicenza found sophisticated ways to express them. The reviewers of a book on Negretti's extraordinary case concluded their commentary in the *Novelle della repubblica letteraria* with a Latin quote from Horace that subtly insinuated that the physician who observed the persistent sleepwalking had deluded at least himself. One needed to be able to place the one Latin sentence in the context of Horace's poem to realize that the reviewers were making sarcastic remarks about the credibility of the physician. Nonetheless, the reviewers' subtle skepticism offers at least a glimpse on the variety of readers' responses to extraordinary medical cases and hands a precious key for reading the primary sources between the lines. If we cannot conclude that Sale and his entourage embellished the case and the cure of the sleepwalker in their publications, we can at least be sure that some eighteenth-century readers believed just that.[94]

As Nollet was well aware, the world of experimental philosophy was not immune from deception and the fabrication of news. The community of academic electricians was not exempt from the charlatanry of the learned that Johann Burckhardt Mencke had so eloquently exposed a few decades earlier. In the description of his experiment called "the beatification," for example, Bose had deliberately concealed a crucial detail, intentionally stirring a controversy that he anticipated would attract international attention to his work. Several experimenters unsuccessfully tried to obtain the effect Bose described, and their repeated failures brought him the fame he expected. When Bose revealed his trick, many complained about his abuse of trust, yet the clamor of the alleged phenomenon, and even more the ensuing scandal, secured his celebrity.[95]

It is no surprise, then, that *Dell'electricismo* foregrounded the fine line between deception and experimental knowledge in the "philosophical and gallant novella" that prefaced the book. The novella was a moral tale that unfolded in the Veneto during the early years of the War of the Austrian Succession—a context that many Venetian readers would recognize as their own. It features two travelers that recall Wabst and Boissaert: a Saxon colonel and a volunteer to the Austrian troops on their way to Venice for the carnival. The story tells their various adventures, leading to the discovery of a lost manuscript, which is *Dell'electricismo* itself. Central in the narration is the sensational tale, which the travelers expose as a deception, of a demon's nightly visits to a local married woman. Using their knowledge of experimental philosophy, the two travelers reveal that the demon is none other than a young apothecary, who uses pyrotechnics and phosphorus to scare the woman's husband away and sleep with her.

The theme of the naïve ignorant, in this case the husband, easily deceived by the cunning artisan was a classic motif in the presentation of experimental science as a subject for the enlightened. Knowledge of pyrotechnics, with which the electric fire was commonly associated, had often been used to articulate this dichotomy.[96] In the Venetian context, in particular, the episode acquired an obvious meaning. In the city, entertaining shows staged by mountebanks, along with miraculous potions advertised by itinerant quacks, were everyday business.[97] Through the story of the cunning apothecary, the book's author humorously reminded his readers that ignorance made them prone to deception. By contrast, electrical demonstrators offered rational recreations that led audiences on a path of self-improvement and enlightenment.

This approach to experimental knowledge as a weapon against deception also characterized the work that Wabst undertook after he went back to Austria. As the physician to empress Maria Theresa, he joined a group of experts who carried out research on vampires in Silesia. The final report on the topic, *Vampyrismus*, labeled a belief in vampires as a superstitious one and set out the chemical reasons that made it possible, under certain conditions, for corpses not to decompose even many years after burial.[98]

Deception was very much in the mind of Nollet when the news of the bishop of Sebenico's cure provoked an uproar in several European countries. In 1748, Johann Heinrich Winkler, professor of experimental physics at the University of Leipzig, declared during a crowded public lecture also attended by government representatives that he had been able to replicate Pivati's experiments. Winkler communicated his results to the Royal Society of London and to the Paris Academy of Sciences.[99] Winkler's report, the first confirmation of Pivati's claims outside Italy, particularly annoyed Nollet, who confided to his friend Jallabert his belief that if Winkler had "no intention of deceiving others, . . . he has no trouble in deceiving himself."[100]

The proliferation of improvised electricians and the risk they presented to the authority of academic experts prompted Nollet to propose that the Paris Academy of Sciences appoint a special committee tasked with performing a new review of all known electrical experiments. The Academy approved the program and mandated that the experiments be carried out at Nollet's home with his instruments.[101] As part of the program, Nollet also obtained authorization from the minister of war, count René Louis d'Argenson, to resume testing electricity with Morand and Bouquot on three paralytic soldiers at the

Hôpital des Invalides. Yet they stopped the electric tests after only fifteen days, because they had failed "to see or notice any other progress that might gratify [the patients'] patience (something most necessary to submit to this sort of torture)."[102] The results were once again inconclusive.

Nollet presented his experiments publicly in the presence of the minister of war and a "large number of people of distinction."[103] Owing to these activities, which he subsequently published, he became the foremost expert on all matters electrical, who demonstrated a distinctive commitment to engage in philosophical duels against deception and self-deception. By 1749, virtually no one would doubt that putting an end to the controversy over the medicated tubes could move him to cross the Alps.

CHAPTER THREE

Fabricated Controversy

A few weeks after his return in Paris, the abbé Nollet presented the first and longest part of his report on his nine-month journey through Italy to the Paris Academy of Sciences. As we have seen, he had dedicated most of his time to gathering information on Italy's silk industry on behalf of the French Bureau of Commerce, but he told his fellow academicians and repeated the claim in print that the main purpose of his journey was "to see for myself the singular effects attributed for some years now to the electrical virtue." Drawing on travel literature that described Italy as a land of marvels, in his "Expériences et observations en différens endroits d'Italie" (Experiments and observations in different parts of Italy), he presented himself as a lover of truth, who had traveled all the way through the Italian states to understand why "all these marvels were as if reserved for a single country." In Turin, "nothing was dearer to my heart," he declared, "than to visit Mr. Bianchi . . . and ask for the experiments . . . that had come out so badly in Paris to be repeated among us and under his direction." In Venice, too, Nollet explained, "one of my foremost concerns was to seek out acquaintances or friends who might introduce me to Mr. Pivati . . . that he might be willing to satisfy my urgent wish to watch him by means of electrification make odors pass through well-sealed glass vessels or noticeably reduce in size some drugs previously sealed inside them."[1]

The confrontation with his Italian counterparts yielded clear conclusions. Affected by a love of the marvelous typical of their land, the Italians had made errors in executing the experiments and had been hasty in their endorsement of the medicated tubes. Nollet presented his journey as a philosophical duel in which the love of truth triumphed over the love of the marvelous. He clearly stated the stakes of the controversy: "If there should be anyone on whom the love of the marvelous can make a victorious impression, I shall not think ill of them, if they embrace opinions opposite to mine. *Qui vult decipi, decipiatur* [those who want to be deceived, let them be deceived]."[2]

Self-deception was the opposite of enlightenment. The philosophical duel over the medicated tubes marked the difference between the modern love of truth, based on the replicability of experimental results, and an outdated love of the marvelous that ought to be eradicated from experimental philosophy.[3] Marvels and wonders had by no means disappeared from the world of natural philosophy: experimental philosophers, including Nollet, often employed the vocabulary of wonder to describe the surprising effects of electrical experiments.[4] Yet principled statements on the printed page contributed to the construction of an imagined reality for experimental philosophy in which ideals of conduct and epistemological dictates did not necessarily correspond to lived practice.

This chapter shows that the philosophical duel was waged only on the printed page. Nollet carefully manufactured his account of his Italian journey and attended to its circulation. Aware of the three-year gap between the submission of reports to the Paris Academy of Sciences and publication in its official journal, he also sent his account to the Royal Society of London, where William Watson, its leading electrician, promptly translated it into English. Nollet's report was published in the *Philosophical Transactions of the Royal Society* in 1750 and immediately reached a vast audience, with several English-language magazines celebrating Nollet's journey to Italy as a philosophical duel against the love of the marvelous.[5]

Nollet's travel diary, however, reveals that the confrontation with Bianchi, Pivati, and Veratti took place over a very short span of time. In Turin, the dispute with Bianchi was resolved over four days, while a single afternoon in Venice was enough to prove Pivati wrong. In Bologna, there was no real showdown at all between Veratti and Nollet. Unpublished documents demonstrate that Nollet in fact was eager to build long-lasting relationships with the Italian scientific community and actively sought the support of Italian colleagues. The publication of Nollet's version of his encounters with the Italian protagonists of the controversy featured a fabricated story that enjoyed fame because of the lessons it offered on the danger of deception and self-deception, and the role of experimental philosophy in the self-fashioning of the enlightened.

The growing popularity of recently created news media, such as gazettes, magazines, and journals, caused an avalanche effect of information that put Nollet's duel in the spotlight. Nollet, meanwhile, hastened to prepare a second edition of his *Essai sur l'électricité des corps* to which he added a section on his Italian journey. Reviewers and readers saluted the brave challenge of the abbé who had traveled to Italy to see with his own eyes how his rivals conducted experiments that only seemed to work south of the Alps. By opposing the love of truth to the Italians' love of the marvelous, they remarked, Nollet won a philosophical duel fought according to shared experimental protocols: "We can take up arms for the truth, we can fight up to a certain point; but when the battle is over we should behave like our warriors during an armistice: they observe each other, they stay on guard, and they say to each other: *now take your seat at the table of allies.*"[6]

Nollet's presentation of his encounters with his Italian counterparts as a philosophical duel was particularly compelling to reading audiences that had learned to conceive of themselves as an enlightened tribunal of public opinion and reveled in tales of disputes and controversies. As a model of how to close experimental controversies, the story of the philosophical duel between Nollet and the Italian electricians met with success.[7] In addition to journals and magazines, it made an appearance in Diderot and d'Alembert's *Encyclopédie*. Decades later, when the medicated tubes were no longer fresh news, authors who wrote on electricity still referred to this episode. François Rozier in his 1777 *Introduction aux observations de physique* (Introduction to observations on physics) and Joseph Priestley in his 1767 *History and Present State of Electricity* both mentioned Pivati's medicated tubes and Nollet's challenge. As it bounced from one page to the next, the philosophical duel became *the* story of Nollet's journey to Italy.

Even the very subject of the controversy, the prodigious cures performed by the medicated tubes, enjoyed a more robust life on the printed page than in any Italian city. Pivati published an embellished version of his experiments modeled on the sensational fake news published by contemporary magazines in search of new readers. False information that got published circulated quickly and, in the minds of many readers, acquired the status of well-established facts. It proved hard to rectify, amend, or retract false information, but more importantly, the very debate about the presumed reality of a story or an event had an extraordinary power to attract the public's attention.[8] Controversies were terrific publicity, and Pivati had reasons for seeking public recognition. This chapter discusses both Pivati's and Nollet's manipulation of information on the printed page.

Gianfrancesco Pivati and the Venetian Book Trade

Although virtually unknown abroad, Pivati was a celebrity in the Veneto. He was a lawyer by training and a member of the local Accademia dei Ricovrati (Academy of the Sheltered). He held the position of superintendent to the book trade for the Republic of Venice, and also served as a censor, charged with the task of examining the contents of books to determine whether they

could be published. He used his official roles to establish connections with local publishers and with the academic community. In 1738, he promoted a
reform of the University of Padua aimed at increasing the teaching of the sciences. Specifically, he identified the chair of experimental philosophy as the
"most useful and most essential to be established."[9] One of his connections and
correspondents, Giovanni Poleni, obtained this chair the following year. His
appointment to superintendent to the book trade put Pivati into close contact
with a major industry in Venice. The superintendent's task was to inspect
printers, check that all the processes complied with Venetian laws, and act as
a spokesperson for the needs of printers.[10]

Aware of the growing popularity of dictionaries, Pivati collaborated with
the Venetian printer Gasparo Baseggio on an encyclopedic dictionary that
would include both sacred history and the newest scientific discoveries. The
Venetian Senate conferred privileges on printers who dared to venture into the
publication of such demanding and costly works, and several other printers
were investing in translations of foreign dictionaries. Among influential
supporters of Pivati's *Nuovo dizionario scientifico e curioso sacro-profano*
(New dictionary, scientific and curious, sacred-secular) were Poleni and his
colleague Giovanni Morgagni, a professor of anatomy at the University of
Padua.[11]

The dictionary's first volume was published in 1746, the same year in which
the electric craze spread in Venice and all over Italy. During the previous five
years, Pivati had consolidated his relations with the Venetian aristocracy and
with doctors, lawyers, and representatives of the clergy and the academic
world, who became subscribers to the work. Since ambitious projects such as
his carried significant financial risks, Pivati had also sought the support and
protection of scientific institutions beyond the boundaries of the Republic of
Venice.[12] In 1746, Pivati sent copies of the first volume of his dictionary to
Francesco Maria Zanotti, the secretary of the Bologna Academy of Sciences.
In the cover letter, Pivati mentioned his relations with Poleni and Morgagni
and asked Zanotti to keep him abreast of the experimental activities of the
academy's members so that he could include them in his work. The maneu-
ver resulted in the election of Pivati to the Academy.[13]

The fake news of the cures with the medicated tubes was part of Pivati's
promotional campaign for his encyclopedic dictionary (fig. 3.1). Like other
electric parvenus, Pivati used the novelty of the field to carve out a space for
himself in the fashionable world of experimental philosophy. He attended
Wabst's and Bossaert's performances and engaged in experimentation in the

Fig. 3.1. This plate from Giovanni Francesco Pivati's *Dizionario scientifico e curioso sacro-profano* (1746–51) shows the "inflaming of spirits" experiment. Bayerische Staatsbibliothek, Munich.

hope of making a breakthrough discovery.[14] The medical applications of electricity seemed the easiest way to demonstrate his dedication to the public good.

His proximity to the book trade gave Pivati opportunities to witness the power of the printed page to make or break reputations. In his capacity as the superintendent to the book trade, he examined charlatans' pamphlets boasting of prodigious cures. Their success demonstrated that exaggeration—if not outright deception—delivered returns. In similarly fashion, in the article on electricity published in his *Nuovo dizionario scientifico e curioso sacro-profano*, Pivati made a number of extraordinary claims about his electrical experiments. He recounted that the scent of Peruvian balsam, which he had placed inside sealed glass tubes, spread all around the air upon electrification. Even more surprisingly, a man who was holding such tubes during electrification sweated copiously during the night and "his sweat, his clothes, and the whole room gave off the vigorous, agreeable scent" of the balsam. The smell, Pivati stated, was left "in his fingers" after he combed his hair and "even lodged in the comb," and "when his clothes, soaked in sweat, were dried by the fire, they continued to transmit the scent all the more." This led Pivati to think that the medicinal properties of certain substances placed inside the electrified tubes could also diffuse in the air and penetrate inside a patient's body through the pores of the skin or even by simple inhalation.[15]

Pivati eagerly looked for supporters of his alleged discovery. In 1746, he sent a letter to Zanotti in which he greatly exaggerated his experimental results and presented himself as the inventor of "medical electricity," *the* momentous discovery that would bring Italy back to the forefront of experimental philosophy. Such vision struck the right key for Zanotti, who could not contain his enthusiasm: "What news in our electrical experiments! What precision! What success! . . . We have proved them in such a fashion that we do not think we are guessing, and as for me, I believe that we have surpassed all others in matters electrical."[16]

Zanotti rushed to publish Pivati's letter in Bologna, although the title page falsely stated it was published in Lucca.[17] Writing under the guise of an anonymous printer, Zanotti explained in the preface to *Della elettricità medica lettera del chiarissimo Signore Giofrancesco Pivati* (On medical electricity: Letter from the illustrious Gio. Francesco Pivati) that "even if electricity has been discussed by others, it can nonetheless be considered new, because it now shows a value that it did not have before, that is to bring a marvelous improvement to medicine, and to offer a ready and easy remedy to many ailments."[18] By exploiting the Bologna Academy of Sciences' network of foreign correspondents, Zanotti undertook to disseminate the extraordinary news and promote the medicated tubes abroad. Pivati's invention, he wrote to Nollet, had "made Italy electric."[19]

In tune with the widespread hope of resuscitating Italy's status in the Republic of Letters, largely perceived as having declined after the golden age of Galileo Galilei and the reputable Accademia del Cimento (Academy of Experimentation), Zanotti emphasized the "Italianness" of Pivati's discovery. As he underscored in his communication to Francesco Algarotti, notoriously fond of all things foreign, "if the facts are true, the Venetian has outdone all those British, French and Germans of yours."[20] Upon learning of the bishop of Sebenico's "surprising" cure from Zanotti, Jallabert hastened to inform the Paris Academy of Sciences about it.[21] Zanotti was not alone in seeing the medicated tubes as an opportunity to relaunch Italian science on the international scene. Iacopo Bartolomeo Beccari, the Bolognese chemist whose interest in electricity dated back to the early years of the eighteenth century, sent a report on Pivati's new method to the French naturalist René Réaumur, who read it during a session of the Paris Academy of Sciences in 1748.[22]

Driven by the ambition to see the Bologna Academy of Sciences ascend to the heights of the most prestigious foreign institutions, Zanotti censored any opposition to the medicated tubes. In his role as the secretary, he had full control over the institution's official publication: *De Bononiensi scientiarum et artium instituto atque academiae commentarii* (Commentaries on the Bologna Institute and Academy of Arts and Sciences). He chose which reports to publish and wrote the introduction to each volume. When it came to formalizing the Bologna Academy of Sciences' response to Pivati's experiments, Zanotti shushed the work of Goffredo Bonzi, a lecturer of medicine, who was the first person in Bologna to repeat the experiments. Using tubes that Pivati himself had sent specially from Venice, Bonzi had been unable to confirm the experiments. In his official report, Bonzi denied that electrification could make odorous substances evaporate through sealed glass tubes and rejected the possibility that the medicated tubes could have any therapeutic efficacy. Zanotti, however, branded Bonzi as one of those people who, "due to their natural inclination to scorn new designs . . . , have no fear of passing off Mr. Pivati's experiments as ridiculous" and refused to publish his report. Instead, he described and praised the medicated tubes, thanks to which, he stressed, "the whole of Italy was excited."[23]

The publication of Pivati's letter on medical electricity kindled the curiosity and ambitions of Bianchi, who was also a member of the Bologna Academy of Science. When Bianchi learned of Pivati's prodigious cures, he applied himself to a long series of tests and sent the results to Nollet and Pivati.[24] Bianchi developed his own variation on the transmission of odors: the electrical purge. Roughly seventy-five years before, Molière had masterfully parodied the obsession with purgatives, laxatives, and enemas through the character of Argante—the imaginary invalid who was a symbol of the hypochondriacs teeming in society. Scammony, gamboge, and aloe succotrina were commonly used to provoke the desired bowel movements, even if the somewhat unpleasant taste of these substances made the treatments rather unpopular, especially with those who had delicate stomachs. Applying Pivati's idea that electrification diffused the medicinal particles into the surrounding atmosphere, Bianchi put the purgatives directly in the hands of the patient. Insulated from the ground and connected to the electrical machine, the patient absorbed the medicine from the air while an operator made sparks shoot from their bodies. The patient achieved the desired effect without having to ingest anything.[25]

According to Bianchi, a goodly number of "Argantes" who were eager to try out the electric effluvia, including university professors and students of medicine, confirmed that the pieces of scammony, aloe succotrina, or gamboge that they held in their hands during electrification produced the wished-for "evacuations." Electrical purges met with success even among those who had not yet tested them. The Bolognese ambassador to the papal court, Flaminio Scarselli, one such mere observer, declared that he was "greatly pleased with this way of purging without having to take any medications by mouth, which often cause disagreeable upsets of the stomach and the intestines."[26]

Several electricians in Italy and beyond tested the electrical purges and repeated the experiments with the medicated tubes, without success. Other leading experimental philosophers besides the skeptical Watson in London expressed their doubts about the medicated tubes and the transmission of odors. Gerard van Swieten, personal physician to the Austrian empress Maria Theresa, reported that in Vienna Pivati's experiments enjoyed no credit as nobody could replicate them. Bose wrote to Réaumur that he failed to obtain the results announced by the Italians; Jallabert also was unable to confirm Pivati's results.[27] Even Bianchi's colleague at the University of Turin, Francesco Garro, sent Nollet detailed information on the failed tests he had carried out himself "following the rigorous Pivatian method." He revealed that "the scammony test was suggested to Bianchi by a pupil of mine, who had formed a particular hypothesis about the electrical virtue and thought his idea could be confirmed with this experiment."[28] Garro had repeated the experiments, placing fresh flowers, Peruvian balsam, benzoin, and many other scented substances inside the globe of his big electrical machine, but he could never confirm that the perfumed substances passed through the electrified glass. He added that Bianchi's servant confessed that he had confirmed the purge only to please his master.[29] The secretary to the duke of Modena, Pietro Ercole Gherardi, reported to his correspondent Ludovico Muratori that Bonzi's experiments definitely showed that the medicated tubes "have the same effect as that produced by incense on the dead."[30]

Learned colleagues warned Zanotti about Pivati's unreliability and even shared that the famous cure of the bishop of Sebenico proved to be nothing more than a lie. The marquis Scipione Maffei, who was also a member of the Bologna Academy of Sciences, explained to Zanotti that in Venice "Pivati's stories" were "believed to be either false or greatly exaggerated." Maffei reported that he had visited Pivati in Venice and found that his electrical machine was "so faulty that in an entire morning there was nothing, however trivial, that he could show me."[31] The Neapolitan professor Niccolò Bammacaro similarly confirmed that a colleague of his who visited Venice was disappointed by Pivati's failure to show even one of the experiments he had so pompously talked up.[32] Maffei told Zanotti that an acquaintance of his in Venice had by chance bumped into the bishop who "to anyone who congratulated him on seeing him finally free of the long podagra and chiragra . . . responded only with a smile and a few shoulder shrugs."[33]

In the face of growing skepticism, Zanotti doubled down in his support of the "Italian" invention. He arranged for Veratti to conduct experiments on medical electricity, urging him to support the medical efficacy of electricity. Veratti faced additional pressures. Scarselli advised Laura Bassi, as did Zanottti, to not wait too long to publish the experimental results. Since "the material is so beautiful, it caters equally to the curiosity and to the diligence of philosophers and physicians," so "any delay" in publishing the results carried the risk that someone else would publish them first.[34] Veratti did indeed publish his results in a book sponsored by the Bologna Academy of Sciences and dedicated to the Bologna Senate titled *Osservazioni fisico-mediche intorno all'elettricità* (Physical-medical observations concerning electricity). Veratti adopted a diplomatic approach to the controversial matter: he stated that, due to lack of time, he had not been able to replicate Pivati's experiments to his complete satisfaction. Yet he started to offer other kinds of electrical treatments in his medical practice and boasted of his own cures. Veratti's book circulated widely and was translated into French in 1750. Bassi sent several copies to Scarselli for him to distribute to cardinals, other philosophers and booksellers in southern Italy, and, especially, to the pope.[35]

The Bologna Academy of Sciences' official sponsorship emboldened Pivati to claim the status of inventor of medical electricity. In 1749, just a few months before Nollet's visit, Pivati published a new book titled *Riflessioni fisiche sulla medicina elettrica* (Reflections on the physics of medical electricity), which trumpeted a whole new set of extraordinary cures. A patient affected by syphilis, reported Pivati, was treated with a special double glass tube filled with mercury and, immediately after the application, his hands took on a leaden color. The next day the patient showed all the effects of being treated with mercury ointments. To a public of potential subscribers to his encyclopedic dictionary, Pivati misrepresented the electrical treatments of Jallabert in Geneva, Le Cat in Rouen, and Winkler in Leipzig as demonstrations of the international popularity that *his* invention of medical electricity had achieved.

Even those who had failed with the electrical cures, he added, had been compelled to visit him in Venice to study his methods.[36]

Adopting a literary strategy commonly used by charlatans and that twenty-first-century readers accustomed to the proliferation of false information will recognize, Pivati blamed his detractors, accusing them of precisely what they were accusing him of. He described them as "mercenaries" and "jugglers," who conducted bizarre experiments in the public squares for the "amusement of the ignorant rabble" and to the "great shame of poor physics." By representing his critics as charlatans, Pivati presented himself by contrast as a trustworthy scholar, whose efforts—both medical and encyclopedic—had restored Italy's leading place in the international Republic of Letters. The printer Benedetto Milocco added his own note to Pivati's new book to remind readers of "the esteem that the author acquired with a discovery so beneficial to humankind." As the publisher of Pivati's dictionary, Milocco was counting on the controversy over the medicated tubes to generate publicity for the expensive project.[37]

Electrified Bodies, Enlightened Minds

Historical and scholarly discussions of the controversy over the medicated tubes have grouped together Nollet's rivals as the "Italians."[38] Those who worked on medical electricity south of the Alps, however, did not constitute a homogenous group. When he arrived in Italy, Nollet joined forces with local colleagues who were skeptical about the medicated tubes. For several Venetians, in particular, Nollet's arrival in the city supplied the opportunity to stage a conflict for which the medicated tubes represented only the tip of the iceberg. The publication of Pivati's dictionary had ruffled many people's feathers, and his rivals were eager to see his reputation undermined.[39] Nollet met many of them in the casino of the patrician magistrate Angelo Querini.

The casino was a Venetian place of conversation, typically a small apartment away from home, where a group of friends could "live at their ease and in freedom." In the *casini* Venetians discussed society events or the latest literary offerings, talked about politics and art, and cultivated philosophical and scientific interests. Ladies of the nobility often rented them to have a place where "conversations of women," particularly feared by the inquisitors, could take place.[40] Querini's casino was also a meeting place for a group of patricians who openly criticized the senatorial oligarchy, many of whom were interested in science.[41] Nollet met there the doge's grandson, who introduced him to the doge himself to whom Nollet gifted his most recent book on electricity.[42]

Among the other members of Querini's circle were the physicians Giovanni Menini and Gianfortunato Bianchini, who were outraged by the prodigious cures paraded about by Pivati. It was Querini who arranged Nollet's visit with Pivati. Nollet thus had the support of several Venetians when the meeting finally took place.[43]

Rivalries in the book trade were more relevant in the Venetian context than any dispute over the methods of experimental philosophy. The printer Giambattista Pasquali, a friend of Querini's, was eager to see Pivati's reputation destroyed. Pasquali's shop, located in the palace of the English consul Joseph Smith, was an active cultural center and a meeting place for the consul's friends. The association between Pasquali and Smith had resulted in the establishment of a printing house that became the largest importer of foreign books in the Republic of Venice. Smith was, along with Pasquali, a promoter of the Italian translation of Ephraim Chambers's *Cyclopaedia*, and he was hostile to Pivati's dictionary, which was a direct competitor for his translation project.[44]

Pasquali regarded Pivati's dictionary as "a silly book," whose success owed solely to the "usual fortune of these sorts of things." Nollet's presence in Venice offered him an opportunity to counteract the success that Pivati's "prattlings" on electricity had garnered.[45] He hoped that disparaging his electrical experiments and showing publicly that he was the purveyor of false information would also discredit his encyclopedic endeavor. Pasquali coordinated with the electricians in Querini's circle, offering to publish the experiments they designed to attack Pivati. The tiffs between Pivati and Pasquali continued over the years: in the guise of superintendent to the book trade, Pivati censured several of the books published by Pasquali, among them Cesare Beccaria's 1764 *Dei delitti e delle pene* (*An Essay on Crimes and Punishment*).[46] Having already published the Italian translation of Nollet's *Essai sur l'électricité des corps* and the first volumes of *Leçons de physique expérimentale* (*Lectures in Experimental Philosophy*), Pasquali hoped that his personal connection with Nollet would result in more such profitable translations. This proved to be the case. He published several other Italian translations of Nollet's works in the following years.[47]

Just as Nollet fully exploited the social occasions offered by his contacts with the Venetian cultural elite, so the members of Querini's casino seized on the opportunities provided by the visit of this distinguished physicist to gain visibility not only locally but throughout the Republic of Letters. Immediately after Nollet's departure, Bianchini launched an experimental program that consisted in systematically repeating all the experiments described by Pivati. Conducting them at Querini's casino and at the home of Menini, in the presence of noblemen and other physicians, Bianchini struck the hardest blow to Pivati's credibility. His *Saggio d'esperienze intorno alla medicina elettrica fatte in Venezia da alcuni amatori di fisica* (Experiments on medical electricity performed in Venice by some physics amateurs), dedicated to Nollet, was published by Pasquali in 1749 and translated into French the following year. Nollet sent an excerpt to the Royal Society of London, which was published in the *Philosophical Transactions*.[48] In the place that had served as a theater for the seemingly miraculous healing of the bishop of Sebenico, Bianchini and his collaborators dismantled Pivati's claims point by point.

As talk spread about Nollet's arrival and his desire to meet with Pivati and confront him over the medicated tubes, the visits of Venetian electricians to the illustrious Frenchman multiplied. In addition to Pasquali and the Querini circle, Sguario too met Nollet to let him know that the misunderstanding about the transmission of odors originated with Pivati's wife, who handled a bottle filled with Peruvian balsam while Pivati was conducting electrical experiments. The highly fragrant substance diffused into the air without electricity playing any role. Sguario added that he knew several people that Pivati claimed to have cured and that none of these healings had actually taken place. In the published version of the controversy, Nollet did not relay Sguario's testimony, perhaps because of the rumor—conveyed to Nollet by Bassi—that he had appropriated Wabst's work or, more probably, because Sguario's stance on medical electricity was by and large negative.[49] Yet it offered Nollet more solid evidence against Pivati and awareness that many Venetians were on his own side.

In the fabricated reality of the philosophical duel, Nollet was open to being proved wrong. To his readers he explained that he was subjecting his conclusions "to the healthy criticism of people more enlightened than me" and described himself in Italy as animated by "curiosity" and desire "to be instructed" by his Italian rivals.[50] In fact, he had decided the outcome of the philosophical duel well before leaving Paris. As he told Jallabert, he was convinced that the Italian news was "at the very least exaggerated."[51] A few days before his meeting with Bianchi in Turin he was even more explicit: "I'm quite afraid," he related in a letter to Jallabert, "that there is nothing here to purge except the imagination."[52] While maintaining an official façade of impartiality, in his private correspondence from Italy Nollet continued to voice his real opinions. After meeting Pivati, Nollet described him as "a man totally new to physics, little accustomed to doing experiments, somewhat of a lover of the marvelous, who believes with great glibness."[53] A few months later he was more straightforward: "He's a charlatan and an ignoramus who deserves no credit."[54]

There was no less partisanship in the other camp. The three Italian protagonists of the controversy exchanged private impressions about Nollet, agreeing that he was biased against the electric cures. After Nollet's visit, Bianchi wrote to Pivati and Veratti, warning them about the danger of performing experiments in Nollet's presence. The abbé was so prejudiced that he attributed "the purge he had to the cold water he had drunk that day," and was "looking for experiments that are favorable to him."[55] After his confrontation with Nollet, Pivati sent his version of the meeting to Veratti, describing Nollet's attitude as a put-down of Italians: "This famous man is, in the prejudices of his nation, against the Italian nation, and . . . I think I can conjecture that he views us as impostors plain and simple."[56] He concluded with a note of skepticism about the real purpose of Nollet's journey: "If this man seeks with such haste to rush off from his observations, for which he says he came from France expressly, and if in little more than an hour he wants to see and be convinced, I'm afraid he will not do much of anything. If an hour of observation in Turin at Mr. Bianchi's and one with me were enough for him, I figure that he'll spend at least two over there."[57]

In fact, Nollet spent several days in Bologna. As in Venice, in Bologna too Nollet sought the collaboration of his Italian colleagues. He had planned this section of his trip carefully, well before leaving Paris. Nollet was aware that the local Academy of Sciences, an institution protected by the pope, had promoted the medicated tubes in Italy and abroad. Bianchi, Pivati, and Veratti were all members. Nollet successfully maneuvered to obtain membership in the Academy through his friend Jallabert in Geneva: "I am not [a member] of the Academy of Bologna, and there can be no doubt whatsoever that I would be very honored to be one; you would do me a great favor if you could help me."[58]

Once he got to Bologna, Nollet presented himself as a colleague eager to make personal connections with his fellow members. Contrary to what he stated in his printed accounts, he told Zanotti that "electricity is, I can assure you, the last of the motives that have led me to Bologna."[59] For their part, the members of the Academy welcomed Nollet very warmly. He was impressed by their friendship and cordiality, "rare qualities among men who follow the same career."[60] Some members traveled to Bologna from other cities especially to meet him. The professors showed him the cabinet of Ulisse Aldrovandi, the collections of Luigi Ferdinando Marsili, the observatory, the chemistry laboratory that was still under construction, and the anatomical wax models that Nollet found "very beautiful." He was pleased to see instruments that had been purchased from his atelier.[61]

During his informal meetings in Bologna, Nollet realized that "minds are divided," and that the members of the Academy regretted "having gone so far" and "decided to be more cautious in the future."[62] Giuseppe Pozzi, the pope's physician, had assured him that "nothing on this matter will come out of Bologna unless it has been carefully examined."[63] This made it easy for Nollet to sport a nonchalant savoir faire when he met Veratti and Bassi in their home. Celebrated in several European publications, Bassi was a distinguished colleague, with whom Nollet shared an interest in electrical phenomena and in the teaching of experimental philosophy. Unlike Pivati, who had composed a "song in the fashion of Anacreon, proving that amorous women should not study," Nollet had devoted a great deal of attention to women's interest in experimental philosophy.[64] He had commented enthusiastically about the marquise du Châtelet's interest in physics and was eager to add Bassi to the list of celebrities to dedicate his work to.[65] Nollet discussed medical electricity with Veratti over the course of several conversations at the Bassi-Veratti home. He praised their electrical machine, even though he privately noted that the glass tubes that Veratti had used to reproduce Pivati's experiments were poorly sealed. After his visit, Bassi and Nollet exchanged several letters, and Nollet dedicated one of the letters in his 1753 *Lettres sur l'électricité* (Letters on electricity) to her.

Although each side endeavored to demonstrate academic savoir faire, the encounters did generate friction on both sides. Pivati's warnings cautioned Veratti not to accept Nollet's proposal to subject Garro, Sommis, and the abbé himself to the electrical therapies. Nollet made caustic comments in his diary that he censored in the official version: "Veratti apologized, saying that F.[ather] G.[arro] would attribute everything that happened to him to the imagination, that in my case, I was not at the time in a state of solid health, and that he (Sommis) complained occasionally of colic, which did not permit him to undergo this type of testing."[66] Regardless, academic diplomacy required that relations between the Paris and the Bologna academies remain on excellent terms and, to this end, Nollet invited Zanotti to become a corresponding member from Bologna. He also got the library of the Bologna Academy of Sciences to agree to buy the Paris Academy of Sciences' journal on a regular basis.

Nollet's opposition to the medicated tubes was bolstered when he went to Rome. The Bolognese pope Benedict XIV, who was the patron of the Bologna Institute of Sciences of which the Academy was a branch, had not yet endorsed the Academy's support of medical electricity. He was eager to know Nollet's impressions of the institution and granted him a private audience.[67] The pope's doctor had revealed to Nollet that the rushed publication of Veratti's work had made Benedict XIV furious.[68] A few years later the pope recalled his feelings about the electric craze of the time: "In every audience with ministers of princes, in every piece of mail that comes from both sides of the mountains, the only effect we experience is that we feel our blood boil, our head stuffed and our bile agitated."[69] Emboldened by the pope's support, on his return to Paris, Nollet constructed a version of the philosophical duel that erased the Bologna Academy of Sciences' role in the promotion of the medicated tubes. In the official report, Nollet never mentioned that Bianchi and Pivati were members of the Academy, nor was the Academy represented as having played absolutely any role in spreading the news of Pivati's cures.

This erasure of the Academy's role in the promotion of the medicated tubes was advantageous to both parties: the Bolognese institution preserved its reputation on the international scene, and Nollet maintained good relations with the Academy and his patron Benedict XIV. Complying with the tacit rules of academic diplomacy, Nollet gave Zanotti the impression that he was open to feedback regarding his account. He sent an excerpt of his report for Veratti to double check, noting that should any inaccuracies be found, he would "make all the changes" Veratti desired.[70] These gestures were largely ceremonial. Nollet had already sent the report to his colleagues in Paris and to the Royal Society of London, where it had been read two months earlier.[71] Veratti—described by Nollet as "a man of sincerity, entirely opposed to imposture"—did notice various inaccuracies and prepared a long reply, but his corrections were never incorporated into Nollet's final version.[72] Nonetheless, in an official letter, Zanotti communicated the Bologna Academy's change of attitude toward the medicated tubes: "Here we hardly even speak anymore of electricity, whether medical or any other kind. The world tires of speaking at length about the same thing."[73]

In fact, the Academy never stopped promoting research on electricity. A few years later, when Benjamin Franklin's electrical theory took Italy by storm, Veratti became a leading figure in a new electrical controversy on the efficacy of lightning rods. In the 1780s, the Academy was again involved in an international controversy on animal electricity, which pitted Luigi Galvani, who had previously collaborated with Veratti, against Alessandro Volta. The Voltaic battery, invented in 1800—an instrument that ushered in a new, crucial period in the history of electricity and humankind—emerged in some measure out of this controversy. At its origins were the questions raised by the frequent electrifications of the human body, which had come out of the entertaining experiments performed in courts and salons.[74]

The Philosophical Duel That Never Was

In print, Nollet presented his journey through Italy and his confrontation with his Italian counterparts as a philosophical duel carried out according to the protocols of experimental philosophy. One of the main tenets of his approach was that the same causes produced the same effects. Nollet had applied this principle in his systematic experimentation on the effects of electricity on animals and plants discussed in the previous chapter, but the assessment of the medical efficacy of electricity presented new kinds of problems. Numerous spectacular experiments demonstrated that the human body's response to electrification varied unpredictably, depending on the subjects' impressionability, clothing, and, more generally, physical constitution. In the field of medicine, it was well known that the same agents acting on different bodies could produce different effects. Yet medical electricity occupied a liminal space between the world of experimental physics and that of medicine. Because of its reliance on instruments, physicians—with few exceptions—were generally either indifferent or skeptical. Experimenters, for their part, did not possess the medical knowledge or authority to establish therapeutic treatments. This liminal status opened up opportunities for a vast range of self-styled medical electricians with no medical education, whose numerous reports of electrical cures threatened the authority of electrical experts.[75]

Nollet's published version of the controversy over the medicated tubes amounted to a normative account of the proper methods for certifying experimental results and a powerful warning message about the risks of deception and self-deception. He presented the philosophical duel as an in-person confrontation that happened in three stages, each of which had a lesson for the readers. The first stage unfolded in Turin, where Nollet interacted with Bianchi, the inventor of the electric purges. The experiments demonstrated that an inappropriate choice of subjects led to inconclusive results. The tests took place in Nollet's own residence, with the machine that the itinerant demonstrator Francisco Bossaert had built for Garro. Nollet was the first to undergo the electric purge. Standing on an insulating support, with a piece of scammony in his right hand, Nollet touched the rotating globe of the electrical machine with his left hand while Bianchi made sparks fly from his body. The entire experiment lasted fifteen minutes. The same procedure was used to electrify a young man, a young woman, the professor Giambattista Beccaria, a kitchen boy who worked in Bianchi's household, and another servant. Twenty-four hours later, only the kitchen boy and the young man said they were surprised by "rumblings in the belly" and subsequent bowel movements. Nollet explained to his readers that he did not believe their testimonies could be considered trustworthy because the kitchen boy had concealed that he had eaten chicory soup for eight days before the electrification, while the twenty-two-year-old was smitten with such a love of the marvelous that it was "prudent to doubt everything he said." In his private diary Nollet added that the kitchen boy had related that even his wife, with whom he had slept, had been purged.[76]

Nollet presented the "lovers of the marvelous" as unreliable experimental subjects. They stood as an example of the primitive state of mind of those who, unprotected by enlightened reason, deceived themselves. Electrical experiments provoked awe and fear in people from the lower ranks of society, who were easily impressed by things they did not understand. Excited by their being invited to participate in an entertainment generally reserved for the elite, members of the lower classes that underwent electrification were unable to distinguish between the effects of electricity and those brought about by impressionability, fear, or enthusiasm. Their testimony was also of dubious value because they were disinclined to offer an opinion that would not be to their employers' taste. Hence, as Simon Schaffer has emphasized, the selection of subjects should be a "delicate" process; "neither infants nor commoners" should be included "but only people of reason," whose status "leaves me nothing to doubt about the certainty of the facts to which they testify."[77] Nollet observed that a different choice of experimental subjects led to less equivocal results. A doctor, an anatomy demonstrator, a marquis, three professors, a count, two tutors of aristocratic children, and Nollet himself all reported that they noticed no unusual movements of the bowels after electrification.[78]

Nollet's "ideal" subject was the definition of the enlightened man. Rational and not impressionable, he was able to assess the effects of electricity fearlessly and impartially. This was all the more important when the alleged effects could instead be produced by causes other than electricity. Nollet himself, for example, suffered indigestion following the evening chez Bianchi when he underwent electric purges. He took pains to explain to his readers that cold lemonade and radishes had "conspired" in his stomach more than electricity.[79] To leave no room for further doubt, however, Nollet omitted the testimony of his colleague Beccaria, who admitted to having had two bowel movements in a single day, contrary to his usual routine—even though the professor had added that the consistency of the feces led him to believe that it was a random event not due to electricity.[80]

Nollet was also eager to defend himself in advance against the accusation that he approached medical matters with "the reasoning of physics." This was a common challenge for medical electricians, and a particularly pressing one for Nollet who had been criticized in France on these very grounds.[81] During his experiments on paralytics at the Hôpital des Invalides in Paris, Nollet had collaborated with physicians and surgeons precisely so that he would not be charged with feigning expertise in a field that was not his own. In Turin, he explained, it was not easy to find a doctor who was willing to spend time away from his own work to observe patients undergoing electrification. Nollet had therefore interviewed Bianchi's patients, but none of them could confirm that the benefits of the treatments actually lasted. Nollet's report did not discredit Bianchi, who was a university professor in the kingdom where Nollet had served as a royal tutor. Instead, Nollet pointed to the patients themselves as unreliable lovers of the marvelous: they "were filled ahead of time with such extravagant hopes and possessed by a sort of enthusiasm that they said and caused to be written much more about it than what took place."[82]

Although Bianchi's unreliable subjects mostly came from the lower ranks of society, the risk of self-deception was not exclusive to any specific social class. Anyone could fall prey to it. The second stage of the philosophical duel, which took place in Pivati's Venetian villa, underscored just that. As Nollet emphasized in his public lectures, proper education in experimental philosophy was key to avoiding the omnipresent dangers of deception and, even worse, self-deception. Unlike Bianchi, Pivati was neither physician nor electrical expert. He was one of the many self-educated amateurs that worried Nollet for the numerous unfounded claims that they circulated and that undermined the credibility of experimental physics. Nollet's published report characterized him as a self-deceiving dilettante, unable to distinguish between natural phenomena and experimental errors. In the presence of about forty people, Nollet related, Pivati refused to repeat any experiment, candidly declaring that the transmission of odors had only succeeded once or twice. Pivati had also explained that the cylinder with which he had observed the phenomenon was now broken. Bypassing the rules of experimental philosophy,

Pivati made the confrontation one about testimony rather than procedure. Even if the cures could not be replicated at will, he claimed, they had indeed happened.[83]

Nollet's discussion of his confrontation with Pivati emphasized the fine line between deception and self-deception. Pivati might well be a credulous dilettante, but Nollet insinuated to his readers that he could in fact be the kind of "learned charlatan" disparaged in 1706 by Mencke in his *Charlatanry of the Learned*. Nollet related that he made himself available to perform the experiments at other times, but no invitation ever came from Pivati. Avoiding explicit epithets, Nollet portrayed his Venetian rival as one of those Italians whose exaggerated claims resulted in foreign tourists wasting their time searching for marvels where there were none. Just as the superlatives that Italians used to describe mediocre monuments deceived tourists, so Pivati's love of the marvelous had dragged experimental philosophers into a pointless controversy. The final blow to Pivati's credibility came from a letter that the Torinese physician Ignazio Sommis sent to Nollet, who quoted from it in his report. Sommis had undergone electrical treatments at Pivati's place without any beneficial effect. Other physicians in other cities, Sommis reported, had tried the medicated tubes with no results.[84]

The third stage of the philosophical duel, the meeting with Giuseppe Veratti in Bologna, demonstrated that experimental philosophy could rehabilitate a lover of the marvelous. Veratti, whom Nollet described as "enlightened, wise, and full of candor," had initially believed in the medicated tubes but was now more cautious. Veratti performed his electric treatments, described in *Osservazioni fisico-mediche intorno all'elettricità*, with "great wisdom, and with a simplicity that speaks the truth." Nollet explained that he and Veratti agreed that the medicated tubes did not work as expected and that no odor could pass through the pores of glass as a result of electrification.[85] In Bologna, there was no real clash because Veratti's initial love of the marvelous dissipated as a result of his interactions with Nollet. Veratti was a former rival turned ally. Readers could be comforted by the idea that even if the love of the marvelous overtook honest and learned people, the methods of experimental philosophy cast light on the path to redemption and truth.

The philosophical duel carefully fabricated by Nollet only existed on the printed page. Quarrels, controversies, debates—whether literary, philosophical, medical, or legal—filled eighteenth-century publications and were the subject of conversation among learned elites. To a reading public that avidly

consumed literary and philosophical news, Nollet offered a conveniently manufactured account that let them decide the winner. Readers had learned to
think of themselves as an enlightened tribunal and enjoyed debating political
matters, legal cases, literary canons, and experimental controversies. Ever
since the publication of Pierre Corneille's *Le Cid* in 1637, which had prompted
a "pamphlet war," several writers had called on the public to adjudicate authors' choices.[86] Authors such as Voltaire, Kant, and Diderot sought out the
tribunal of public opinion to legitimate their views, and Nollet did the same
in several of his publications. He typically presented both criticism he received
and his response to it as materials for his readers to adjudicate. In the fabricated reality of Nollet's travel to Italy, the controversy over the medicated tubes
concluded itself with an in-person confrontation.

In this fabricated reality, Nollet consolidated also his "print presence" as an
impartial arbiter who brought experimental civility to the land of marvels.
This version sharply contrasts with what his travel diary and the private correspondence among the protagonists of the controversy reveal. Nollet was
eager to build connections and strengthen existing relationships with the
scientific community in Italy. Neither party was ready to change their mind
as a result of experimental trials. Nollet was aware that there was no unani-
mous consensus on the medicated tubes among his Italian colleagues, and he
strategized his alliances well before leaving for Italy so as to build up his reputation among the Italian learned community. He worked together with the
Italian opponents of the medicated tubes and was careful to take steps that
would prevent him from being perceived as a rival, especially among his academic colleagues who had praised his contributions to experimental philosophy and, in many cases, had purchased scientific instruments from his
atelier.

Closure

Nollet's published account did not make the medicated tubes disappear. The
battle waged on the printed page for decades and kept interest in the tubes
alive. After Nollet presented the French translation of Bianchini's *Saggio
d'esperienze intorno alla medicina elettrica fatte in Venezia da alcuni amatori
di fisica* to the Paris Academy of Sciences, a French edition of Pivati's *Della
elettricità medica* was published in France. In 1755, a full six years after Nollet's
trip, Algarotti wrote that despite the fact that electrical purges were reso-
lutely denied by Nollet, whom he sarcastically labeled the "archon in this

province of philosophy," a member of the Berlin Academy of Sciences electrified "five putti, who held aloe succotrina in their hands," and one of them,
"the day after, had three evacuations of the bowels." According to Algarotti, the
experiment was repeated with a larger machine: operators extracted sparks
from two boys who held gamboge in their hands for half an hour. The result
spoke clearly: "they all had bowel movements, for several days after the treatment. A surgeon kept his eye on the food they ate."[87] The fabricated facts that
circulated on the printed page fueled support for self-styled medical electricians, who continued to offer and design new electrical treatments for decades.
Veratti himself never stopped electrifying his patients. The medicated tubes
gradually disappeared as new electrical discoveries transformed the practice
of medical electricity.[88]

The behind-the-scenes negotiations revealed by private correspondence
and Nollet's travel diary show that published documents can trick historians
into taking the fabricated printed stories at face value. What we lose when that
happens is not only historical accuracy but above all the chance to discuss the
public presentation, or misrepresentation, of scientific information as part of
the skillset of the experimental philosopher. Both Nollet and his Italian
counterparts knew that, whatever informal comments were made off the record, in the end what really mattered was their "print presence" in the published
version of the controversy. When Scarselli told Bassi that Nollet had "publicly
accused the alleged purgative experiments of imposture" and had not spared
Veratti, Bassi asked for Scarselli's discretion, explaining that her "utmost concern" was to preserve her good relationship with Nollet.[89] An expert in academic diplomacy, Bassi was interested in positively affecting the version that
Nollet would eventually publish. In the end, the private animosities, the biased
assessments, and the nationalist sentiments would give way to an imaginary
account of a philosophical duel between the love of the marvelous and the love
of truth. The utmost concern of all parties involved was that their reputations
be kept intact.

Accounts of extraordinary ailments cured by electricity challenge modern
readers' expectations and invite questions regarding the connection between
published stories and their historical reality. In the twenty-first century, we
confront the consequences of the proliferation and mass circulation of fake
news. Unsubstantiated statements and doctored images that spread on social
media create alternate realities that induce many people to action. Scientific
articles that are retracted because they contain fabricated data continue to influence public discussions and individual choices on health or environmental

issues.[90] Eighteenth-century readers were similarly called on to make decisions about what to believe and who to trust. The emerging news market included scandalous libels, gossipy booklets, and advertisements of medical panaceas. News writers amplified details or made up stories wholesale that once in print circulated fast, sometimes with catastrophic effects.[91] Printers commonly recycled news published decades or even centuries earlier, passing off past events as new after a quick change of place and date.[92] The very uncertainty about whether the prodigious cures occurred and whether the sicknesses were as described contributed to the fortune of those who published such accounts.

The power of the printed page to turn made-up stories into matters of facts worried many eighteenth-century authors. In 1707, the astronomer Francesco Moneti complained that "the world was plenty with simpletons ready to believe all they read." The very same group with which Pivati collaborated, the Reformers of the University of Padua, which supervised matters related to education and learning, grew increasingly concerned with the circulation of "false facts" and those that "alter the substance of truth." In 1770 the group charged Pivati's successor with monitoring the circulation of fake news.[93]

The long-lasting fame of the fabricated philosophical duel over the medicated tubes reveals that enlightened experimenters considered the "love of the marvelous" a real threat, worthy of being battled. They claimed that reliable knowledge was to be achieved by means of a self-evident truth emerging from experiments that anyone should be able to replicate, but they lived in a world in which readers continued to be fascinated by mermaids, monsters, and extraordinary medical cases.[94] In this world, deceptions passed off by the vulgar—for example, the "poor woman" Mary Tofts who tricked physicians into believing she gave birth to rabbits—coexisted with those passed off by the learned. In 1699, the *Nouvelles de la république des lettres* (News from the Republic of Letters) published the news of an incredible discovery by François de Plantades, a naturalist and a founding member of the Montpellier Academy of Sciences. Under the pseudonym of Dalempatius, de Plantades claimed to have observed fully formed miniature humans in spermatozoa through a microscope. De Plantades intended these claims as a satirical intervention, but many microscopists took the news for real and followed up.[95] It is unsurprising that just a few years later the editor of the *Acta eruditorum* railed against the "charlatanry of the learned," prompting a debate in which several participants called for publicly shaming those who circulated hoaxes or otherwise spread misinformation.[96]

In the mental space created by the printed page, experimenters eager to pass as experts competed for attention with a number of improvised demonstrators who dangerously wobbled on the line between education and deception. The activities of these demonstrators, some of them full-fledged academicians or professors, introduced confusion over what constituted reliable knowledge, casting shadows over the truth claims of experimental philosophers. If, as Lorraine Daston has argued, enlightenment was a process of self-definition, then guarding oneself against deception was an enlightened way of life.[97] In this framework, Nollet's fabricated exposé became for many readers a powerful wake-up call about undetected deceptions and their consequences on science's progress.

It was no coincidence that Priestley resumed the controversy in his *History and Present State of Electricity* almost two decades later. Priestley believed that the role of history was to illuminate the "rise and progress" of natural experimental philosophy. Hence, the historian's task was to create chronologies that illustrated how natural knowledge had developed over time, erasing controversies, debates, or other faux pas from their accounts: "I made a rule to myself, and I think I have constantly adhered to it, to take no notice of the mistakes, misapprehension, and altercations of electricians; except so far as, I apprehended, a knowledge of them might be useful to their successors. All the disputes which have no way contributed to the discovery of truth, I would gladly consign to eternal oblivion."[98]

The history of the sciences, then, ought to be a manufactured story whose goal was to illustrate how progress should be achieved. That Priestley mentioned the controversy over the medicated tubes, even as he declared that he had no interest in the "altercations of electricians," indicates that he believed his readers could still draw a lesson from it. Namely, he wanted them to understand that the best way to combat the ever-present threat of deception was to place trust in the methods and instruments of experimental philosophy. No matter if experimental philosophers themselves deceived their readers with the presentation of a largely fabricated controversy—or of a similarly fabricated history.

CHAPTER FOUR

Natural Marvels, Instruments, and Stereotypes

Italian nature was part of the Grand Tour experience, just like visits to collections, libraries, and archeological sites. Yet, the Italian natural marvels became destinations of their own for travelers who dedicated themselves to natural history. Their nature-focused tours, which scholars have labeled "naturalistic journeys," became increasingly popular toward the end of the eighteenth century.[1] With thermometers, barometers, electrometers, and other scientific instruments in their bags, naturalists like Jean Jallabert, Horace Bénédict de Saussure, and Jérôme de Lalande traveled through Italy animated by the desire to carry out experiments in situ. Naturalistic journeys provided opportunities to build reputations in the emerging field of the earth sciences and extended itineraries to the hard-to-reach Italian South.[2] This increased emphasis on southern destinations reflected also on the routes of late eighteenth-century grand tourists who, more often than their predecessors, traveled through the southernmost part of the Italian peninsula and reached Sicily. Travelers had long regarded Italy as "nature's masterpiece," but for those like Johann Wolfgang von Goethe, reaching Sicily was tantamount to grasping Italy's true essence. "Italy without Sicily leaves no image on the soul: here is the key to all," he wrote from Palermo in 1787. Goethe was enthralled by the numerous ancient ruins as well as by the Sicilian countryside, whose "strange influence" pervaded him "in the most agreeable way possible."[3]

Travel literature often presented the southern destinations as newly discovered lands. In the later eighteenth century, several travelers framed the wilderness they saw in the Italian South in terms that echoed the narratives of the exploration and conquest of the Americas, describing southern Italians as less civilized than northern Europeans.[4] Before setting out on their tour, travelers prepared themselves by reading travel literature and works of natural history. Along with information, they absorbed stereotypes about local people and their beliefs. Scholars have addressed the role of the Grand Tour in creating and circulating stereotypes about Italians, especially the Neapolitans.[5] As Melissa Calaresu has argued, several grand tourists articulated such stereotypes by building on climate-based theories of human difference, whose popularity peaked after 1749, when Montesquieu presented them in his *Spirit of the Laws*.[6] But the stereotyping of Neapolitans had a longer history, dating at least to the sixteenth century, when Pandolfo Collenuccio's *Compendio delle historie del regno di Napoli* (Compendium of the histories of the kingdom of Naples) associated their fiery and rebellious character with the volcanic soil on which they lived.[7]

This chapter explores the connections between the debunking of natural marvels and the construction of stereotypes, by situating Nollet's visit to Naples and environs in the broader context of naturalistic writings about the Italian South. Addressing an audience accustomed to reading travel accounts about Italy, in his published reports Nollet eagerly underscored that his journey was different from the Grand Tour because of the attention he devoted to the observation of natural phenomena: "Since nature in this beautiful part of the world so abounds with phenomena," he explained, "I could not take it into consideration, albeit briefly, without making a few observations that prompted me to reflect and conduct experiments." Offering a striking counterpoint to Daston and Park's view that enlightened philosophers did not engage in debunking marvels, Nollet stated that he intended to offer a new understanding of "what in the public voice is called marvelous."[8]

Although the Italian marvels of nature had been touristic attractions for centuries, the experiments that Nollet carried out on site were consistent with a recent approach that brought together natural history and experimental philosophy.[9] Nollet deployed scientific instruments to debunk celebrated natural marvels and to draw a clear boundary between the enlightened and the lover of the marvelous. I show that Nollet presented the love of the marvelous as a misleading approach that resulted from ignorance rather than from climate. By exposing the locals as victims of self-deception, Nollet's accounts articulated an argument that went beyond his own journey and the specific case of Italy. He turned the battle against the love of the marvelous into an enlightened project toward self-improvement to be carried out with the instruments and methods of experimental philosophy. Nollet showed his readers how to deploy experimental philosophy to debunk false information that had circulated, unchecked, for centuries. The report of his fieldwork displayed enlightenment in action: the acquisition of natural knowledge through tests and experiments liberated naturalists—and with them their readers—from self-imposed ignorance.

This battle took place on the printed page more forcefully than in the real-life interactions with local people. Nollet's unpublished travel diary and private correspondence, much more than his published account, reveal the culture of reciprocity that characterized his real life encounters with the Neapolitan savants. At the same time, Nollet's travel diary points to the strong influence that the stereotypes of Italy as a land of marvels and of the Italians as lovers of the marvelous exerted on him. This is particularly evident in Nollet's description of one of the most spectacular events in Naples: the miracle of the liquefaction of St. Januarius's blood. I argue that more than the miracle itself, what attracted the attention of Nollet and other foreign travelers was the Neapolitan crowd, which in their accounts epitomized all that made Italians so essentially unreliable: when they were not actively deceiving others, they were being deceived themselves.

Naturalists and Stereotypes

The "discovery" of southern destinations introduced new categories for understanding Italy—such as the "picturesque"—and added the vocabulary and the spirit of the voyages of conquest to the well-established genre of the *voyage d'Italie*.[10] Travel accounts emerging from naturalistic journeys to the Italian South often included descriptions of local people and shared narrative strategies with the ethnographies of extra-European places published in the very same years.[11] These accounts created imagined realities in which southern Italy was the frontier of European civilization, more similar to the American colonies than to Europe itself.

Travel writers built on a tradition dating back to the sixteenth century, which presented the kingdom of Naples (extending to Sicily) as the Italian version of the West Indies. Adriano Prosperi has demonstrated that the idea of southern Italy as the "other Indies" originated with the conquest of Naples by the Spanish crown in the sixteenth century and the Jesuits' need to justify their evangelical missions. Jesuit missionaries described people in southern Italy in the same terms they used to characterize the indigenous people they encountered in America: they were childish, uncivilized, irrational, lawless, and primitive. The similarities between the Italian South and the West Indies extended to the natural world. The sixteenth-century Spanish chronicler Gonzalo Fernández Oviedo compared the natural environments in Sicily and in Central America, representing the kingdom of Naples as the Old World version of the West Indies. Two centuries later, the French academician Charles de La Condamine used his knowledge of South America to describe the Italian mountains, highlighting similar features between the two places.[12]

The most influential text that othered southern Italians was Patrick Brydone's *A Tour through Sicily and Malta* (1773), one of the ten best-selling books in late eighteenth-century England and a travel companion to many grand tourists, including Goethe. Brydone toured Italy with Jallabert and de Saussure and described at length the experiments the group conducted in remote locations. He guided readers through an extraordinary and relatively unknown landscape, introducing them to the methods naturalists followed to measure geophysical parameters, such as the height of the mountains, the variation of pressure with altitude, and the composition of the air. Brydone also referred to the locals living in the environs of Mt. Etna in ways that could have come straight out of the pages of the conquistadores. "We have found a degree of wildness and ferocity in the inhabitants of this mountain, that I have not observed any where else." He also explained that it was no surprise because "where the air is most strongly impregnated with sulphur and hot exhalations . . . people [are] always most wicked and vicious." Brydone further characterized the inhabitants of the areas around Etna as greedy simpletons, ready to believe any sort of story and to fool the tourists at the earliest opportunity. At the sight of Brydone's scientific instruments, they ran away in fright.[13]

These were all well-rehearsed motifs in European narratives of cross-cultural encounters, which typically represented science and technology as hallmarks of the European modern way of knowing that stood in opposition to the superstitious beliefs and primitive habits of local simpletons. Jean-Jacques Rousseau was convinced that "the Europeans have, in consequence of their arts, always passed for Gods among the Barbarians": among "ignorant people," he clarified, "such instruments as the Cannon, the Loadstone, the Barometer, and Optical Instruments" could work all sorts of "prodigies."[14] This notion had a visual equivalent in the engraved silver plate that the Jesuit friar and professor of philosophy in the Maynas mission in South America donated to La Condamine during the latter's visit to Peru. The plate shows Minerva, the goddess of knowledge and personification of the Paris Academy of Sciences, together with several putti that are operating scientific instruments. The background is divided in two sections. To the left, France is represented through symbols of scientific and technological modernity: the Paris observatory, a telescope, a clock, an astronomical sextant, a cannon with military compass, a surveying instrument, a globe, and a compass. To the right, South

America is represented through an uncultivated landscape that features a palm tree in the center and a Spanish colonial structure to the right. In this section putti represent the moderns' approach to indigenous nature: they engage in microscopic observations of local flora and fauna and in chemical analysis of local minerals. In other words, scientific instruments bestowed upon the moderns the power to study, understand, and evaluate local resources (fig. 4.1).

The visual vocabulary that coded scientific instruments as symbols of European modernity had an illustrious precedent in Johannes Stradanus's circa 1590 collection *Nova reperta* (New inventions). One print in the series, intended as an allegory of America, represents Amerigo Vespucci as he encounters America, who is portrayed as a naked woman with a feathered headdress and skirt sitting on a hammock in a wilderness that includes a scene of cannibalism. The explorer, by contrast, is depicted near two large ships that evoke marine commerce and explorations. He wears rich clothes, sports a sword, and holds a mariner's astrolabe in his hand, a navigational instrument that stands between him and America as a symbol of modernity and superiority (fig. 4.2).[15]

Fig. 4.1. This plate from Charles-Marie de La Condamine's *Journal du voyage fait pour ordre du roi à l'equateur* (1751) represents French science reaching South America. John Carter Brown Library, Providence, RI.

Fig. 4.2. Johannes Stradanus, "Allegory of America," from *Nova reperta* (ca. 1590). Metropolitan Museum of Art, Elisha Whittelsey Collection, New York.

Simon Schaffer has argued that reaching distant places was for European explorers a sort of time travel through the ages of humanity.[16] Reaching Italy, and in particular its southernmost areas, was for many travelers a similar experience, connecting them to the Greek and Roman roots of European civilization. In 1751, La Condamine described the magical feeling of traveling back to the "beautiful days of Greece" when he saw a literary academy gathering inside an amphitheater in Rome. For a fleeting moment, he imagined himself as an eyewitness to the distribution of medals "to the winners of the Olympic games." The tale of his delightful daydreaming, however, contrasted sharply with his account of the Italians' "visibly barbaric" customs of measuring time according to a method based on "vulgar ignorance" of astronomy.[17] The Italian twelve-hour day, explained La Condamine, started at sunset, a moment that varied daily and even more visibly depending on the season. This meant, he clarified, that in a city like Rome, at the moment of the astronomical noon (when the sun is at its highest position in the sky), clocks struck the sixteenth hour on the day of the summer solstice, whereas they struck the nineteenth hour on the day of the winter solstice. Furthermore, because of the earth's rotation, clocks needed to be adjusted periodically with the aid of tables. This way of measuring time made it impossible for Italians to properly set watches or even to realize that a watch did not work. As Schaffer has argued, the proper measure of time was a disciplining process for the soul.[18] It was no surprise, then, for La Condamine and his readers that Italians should be lovers of the marvelous or otherwise morally or intellectually unruly. To further emphasize what he regarded as the Italians' backwardness, La Condamine underscored that in a city as important as Rome there were many literary academies but not even one dedicated to the sciences.[19]

Natural Marvels for Tourists

Travelers' anticipated itineraries had to take into account the morphology of the Italian territory. Long stretches of unpaved roads between cities or the crossing of the Apennines on the way to Florence required careful planning. The availability of lodgings along the routes of long journeys, in particular, often determined the final itineraries and turned remote locations into touristic attractions. The constant flux of visitors, whether grand tourists or other kinds of travelers, promoted the development of an emerging tourism industry, well documented in travel guides such as Misson's *New Voyage to Italy*. The various taverns and hotels that catered to travelers on tour sprung up near the most popular destinations.[20] At the same time, the reputation of lesser-known places often grew in tandem with the presence of a nearby *locanda*. These natural sites shaped the visitors' appreciation of Italy no less than its artistic heritage.[21]

Locals quickly turned tourists' interest in natural marvels into business opportunities and actively contributed to the representation of Italy as a land of marvels. They mediated tourists' experience of the places, facilitating access to remote locations and, in some cases, staging spectacular demonstrations. Their scripted presentations of local marvels reinforced stereotypes circulated by travel literature—often intentionally. Some touristic destinations like the Cave of the Dog owed their popularity to the local guides. The Cave of the Dog was a small natural cavity near Lake Agnano, not far from Naples, which took its name from the demonstration that the local guard routinely performed for tourists (fig. 4.3). The guard forced a dog to breath near the ground inside the cave, showing that it would go into convulsions and quickly die unless promptly taken to the lake close by. The Cave of the Dog had attracted tourists

Fig. 4.3. Tourists at the mouth of the Cave of the Dog ready to witness the guard's demonstration. This print appeared in the 1722 edition of Maximilien Misson's *Nouveau Voyage d'Italie*, when the growing interest in Italy's natural marvels supported more expensive publishing ventures. Wellcome Collection, London.

for centuries. Travel guides entertained tourists with tales of famous visitors forcing dogs, birds, reptiles, and even servants to breathe in the mysterious vapor that circulated inside the cave.

These stories then appeared on the printed page and spread widely even among those who never traveled. Misson's readers learned that Charles VIII, who visited the cave after conquering the kingdom of Naples in the fourteenth century, requested the demonstration on a donkey, while the viceroy Pedro from Toledo forced two enslaved people to breathe close to the ground until they both died. A man by the name of Tournon bent down inside the cave to pick up a stone and he, too, died shortly afterward, despite being carried to breathe near the lake.[22] These anecdotes, repeated almost in exactly the same terms in travel books, sustained the imagined reality of Italy as a land of marvels.

Local naturalists, too, profited from tourists' interest in natural marvels, feeding their curiosity and expectations. They accompanied travelers to places

of interest, where they offered lectures and performed measurements or demonstrations that emphasized the uniqueness of the place. Unlike local guides,
naturalists did not get monetary compensation for their activities. Their returns consisted in new connections that boosted their reputations and in the
possibility of seeing their name in influential publications. Local naturalists
were aware of the tourists' interest in the most recent archeological findings
and adapted their guided tours accordingly. In 1755, a few years after the ex-
cavation of the Temple of Serapis in Pozzuoli, Neapolitan professor Giovanni
Maria Della Torre, who organized periodic excursions to the top of Vesuvius,
published *Storia e fenomeni del Vesuvio* (History and phenomena of Vesuvius),
a book that showcased his expertise on the matter. It presented detailed plates,
engraved from drawings taken with a camera obscura. One of the plates represents a lava formation that Della Torre wanted his readers to see as a "large
grotto in the shape of a round temple," the same shape as the Temple of Serapis (fig. 4.4).

This presentation drew on the Renaissance notion of *lusus nature* (jokes
of nature) and of "nature as artist" that in Naples had illustrious prece-
dents. In his 1599 *Dell'historia naturale* (On natural history), the famous

Fig. 4.4. A plate in Della Torre's *Storia e fenomeni del Vesuvio.* Its caption reads
"large grotto in the shape of a round temple formed by lava in Ottaviano." Beinecke
Rare Book and Manuscript Library, Yale University, New Haven, CT.

sixteenth-century Neapolitan collector Ferrante Imperato described various "figurated stones" in his collection, natural minerals or fossils whose shape or appearance seemed to suggest that an artist (nature itself) had engraved or sculpted them. The figurated stones were natural marvels, and Della Torre's choice to present the lava formation in this tradition indicates his participation in the construction of the area as marvelous. Marvels were extraordinary and unique, and for Della Torre they constituted opportunities both to lure learned tourists to Vesuvius and to create a special niche for himself in the field of the natural sciences. His approach to studying the area around Vesuvius was predicated on the notion that local phenomena could not be reduced to universal laws. Their study required repeated observations, measurements, and experiments that gave locals an edge over any foreigner who could carry out research only for a short time. Mentioned by various foreigners in their publications, Della Torre also encouraged local authorities to take an interests in the region's natural history. For example, Cardinal Spinelli, who had a vacation palace at the foot of Vesuvius, routinely joined the excursions and became Della Torre's powerful patron.[23]

Such authorities were eager for tourists to experience Naples and its surroundings as a place of wonder. They built a dedicated route for tourists to reach the volcano, which was neither the easiest nor the shortest among several other possibilities, yet it forced them to pass through the king's palace and a triumphal arch—a symbolic statement of the political power's investment in the area, rich with Roman ruins and mining sites.[24] To reach the crater, tourists rode mules up to a certain point and then had to "use their hands and feet to scramble up." Alternatively, they could "be dragged or pushed by peasants who earn a living by leading the curious in this fashion."[25] Once on top, they appreciated even more the learned conversation and scientific activities of their guides. Back in Naples, they could purchase a souvenir of their brave endeavor from a "lava dealer" or a collector of minerals.

The ideal of Italy as a land of marvels, which travel accounts constructed and spread, was reinforced by locals and exerted a powerful influence on eighteenth-century naturalists on tour. While tourists contented themselves with admiring natural phenomena and demonstrations staged by local guides, traveling naturalists sought new explanations for the marvels of nature that they read about in travel accounts. In the report of his excursions in the surroundings of Naples, read to the Paris Academy of Sciences in 1750 and published four years later, Nollet demonstrated that the instruments of experimental philosophy could demystify allegedly marvelous phenomena. Like other tourists, Nollet prepared for his journey by reading travel guides and naturalistic accounts. He was particularly intrigued by reports concerning the Solfatara, a large volcanic and mining area full of bubbling waters and sandpits emanating sulfuric exhalations that had inspired travel writers since the time of Pliny (fig. 4.5).

Steeped in the tradition of marvelous accounts, stories about the Solfatara had long satisfied readers' appetite for surprising phenomena. Recycling details from previous writers, these works described the extraordinary properties of the area's hot waters, sulphuric vapors, and bubbling sandpits. In 1550, historian Leandro Alberti published *Descrittione di tutta Italia* (Description of Italy), a highly influential work that included numerous anecdotes about the Solfatara. In typical sixteenth-century style, Alberti made ample use of previous works, including those by the celebrated forger Annius of Viterbo. Alberti's book was a major inspiration for the first travel guide authored by a local, Domenico Antonio Parrino's 1725 *Nuova guida dei forestieri per osservare e godere le curiosità più vaghe e più rare della fedelissima gran Napoli* (New guide for foreigners for observing and enjoying the most beautiful and rarest curiosities in the most faithful great Naples), which became essential reading for foreign travelers. Drawing on Alberti, Parrino related that anything introduced into a particular sandpit at the Solfatara would disappear, including "a knight who had the temerity to ride across it with his horse." He also pointed out that the boiling waters of the Solfatara were so hot that people could easily cook eggs in them.[26]

Fig. 4.5. A depiction of the Solfatara as a marvelous site in Sieur de Rogissart's *Les delices de l'Italie* (1707). Yale Center for British Art, New Haven, CT. Gift of Frederick W. Hilles.

In his published report, Nollet explained that the desire to debunk these marvelous stories strongly motivated him: "The more these places are frequented by travelers and celebrated by authors who publish the marvels of Italy, the more I was eager to see them." He emphasized that these marvelous accounts could and did have the dangerous effect of preventing research and the advancement of knowledge. "People of the place," he explained, always told tourists "the story of a knight who was swallowed by this abyss." By doing so, they showed "their fear, and pass[ed] it on to foreigners." However, Nollet related to his readers that his exploration of the area indicated that the abyss that travel guides described so vividly was nowhere to be found.[27] The locals deceived themselves because of their blind belief in stories steeped in the culture of marvelous. They became prey to fear and deception in contrast to enlightened truth seekers who trusted their own senses, intellect, and instruments.

In Nollet's account, a simple instrument like the thermometer helped to distinguish between accurate and false information. A local told him that the Solfatara waters were so hot that they melted organic materials. Whereas most tourists would believe this statement and maybe even repeat the story in their publications, Nollet showed with a thermometer that the temperature of the Solfatara's bubbling waters was much lower than that of water's boiling point. It was therefore impossible to cook eggs in them, contrary to what Alberti and the local people stated, let alone for the waters to melt anything. Similar stories circulated in Tivoli, near Rome. The locals believed that the waters of Acqua Zolfa, a yellowish stream from which a strong sulfur smell emanated and that fed into little lakes with bubbling waters, were dangerously hot, and they generally feared the place. Yet the thermometer once again demonstrated that the temperature was only slightly higher than that of the surrounding air. According to Nollet, the bubbling was due to a mild-temperature vapor issuing from the subsoil. A simple measurement invalidated an idea that been accepted uncritically for centuries.[28]

For Nollet, the love of the marvelous deceived and distracted people from the pursuit of knowledge. Experimental philosophy, by contrast, identified the causes of natural phenomena and indicated how to make sense even of the strangest ones. It was precisely when it confronted such unusual natural phenomena in the field that experimental philosophy demonstrated its power. By explaining away presumed marvels, instruments like the aeolipile, the air pump, and the steam engine reaffirmed the key principle that the operations of nature followed the same laws independently of place: Italy was a land of marvels only for those who wanted to believe so.

Those in search of knowledge and truth relied on the evidence provided by the instruments of experimental philosophy. These instruments were not just measuring tools. They were technologies of knowledge that activated new ways of thinking about unusual natural phenomena. The aeolipile, for example, enabled Nollet to present an explanation of the surprising bubbling of the waters of the Solphatara based on analogy between the device and the area around the crater. The instrument, originally invented by Hero of Alexandria and rediscovered in the sixteenth century with the publication of the works of Vitruvius, demonstrated the propelling powers of expanding vapors. In its most recent version, employed by Nollet in his lectures, the aeolipile consisted of a small pear-shaped metallic vessel filled with water and closed with a cork on its thin side (fig. 4.6). The vessel was placed horizontally on a chariot and heated by fire. When the water in the vessel started boiling, the steam pressure increased and eventually pushed out the cork. The entire machine then started moving in the opposite direction as the steam cooled down.

In his report to the Paris Academy of Sciences, Nollet used the aeolipile as a model to explain that the bubbling of the waters at the Solphatara was caused by a stream of underground vapor that became cooler as it expanded and reached the surface. The vapor made the waters it passed through bubble, yet in its trajectory upward, it also became less hot, in line with his temperature measurements. By exposing the fallacy of Alberti's and Parrino's marvelous accounts, Nollet demonstrated that even learned readers could fall prey to self-deception when overcome by the love of the marvelous. The instruments of experimental philosophy constituted the weapons of choice for a victorious fight against the dangerous threat of self-deception.[29]

In Nollet's account, philosophical instruments were tools of enlightenment. Not only could they debunk presumed marvels, exposing centuries of ignorance and deception: they also offered new ways of making sense of unusual natural phenomena. Nollet presented his excursions to the crater of Vesuvius and to other natural marvels in the Neapolitan countryside in this spirit. He demonstrated that experimental practice and knowledge of the instruments of experimental philosophy could lead to theoretical breakthroughs. Although the activity of the volcano at the time of his visit made it impossible for him to carry out as many experiments as he had wished, he wanted his readers to understand that the phenomena observed on top of Vesuvius were similar to the violent explosions provoked by the incorrect handling of steam engines, such as the fire pump, the aeolipile, or Papin's machine. Based on the analogy between these machines and the phenomena he observed on the volcano, he

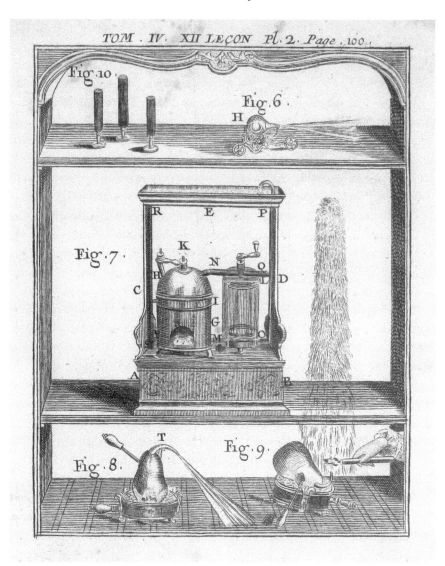

Fig. 4.6. A plate from Jean-Antoine Nollet's *Leçons de physique expèrimentale* (1745) representing the aeolipile (fig. 6) and a fire pump (fig. 7). Musée d'histoire des sciences, Geneva.

advanced the hypothesis that Vesuvius acted like a fire pump: it sucked up water from the sea, heated up the water in its interiors by means of the earth's subterraneous fire, and eventually threw the steam and everything that obstructed its flow violently upward when the pressure became so intense as to break the layers of rock that covered its mouth. Anyone who attended

Nollet's course of experimental physics could agree on the plausibility of the explanation.[30]

As he rewrote the narrative about the Italian marvels of nature, Nollet demonstrated the power of experimental philosophy to help guard oneself from self-deception. His experiments at the Cave of the Dog debunked the common notion that the waters of the nearby lake had special vivifying qualities. Nollet investigated the widespread idea that the vapor inside the cave was poisonous by performing routine tests typically used for investigating natural substances, particularly vapors. The air pump, a classic instrument of natural philosophy, was essential reference for anything related to respiration. Nollet drew on the repertoire of experiments carried out with this device to study the nature of the vapor in the cave. He inserted a candle flame in the grotto and noted that the vapor extinguished it. He then examined the reactions of insects, worms, and small reptiles and observed that animals that had been immersed in the cave returned to their senses completely after breathing outside for a few minutes. All these observations convinced him that the vapor could not be a kind of poison. He further tested the effects of the vapor on his skin or eyes; he noticed no reaction, and so he decided to enter the cave to inhale the air for a few seconds. The only remarkable effect was a slight sensation of suffocation, which soon passed as he walked out of the cave. He had no need to go to the lake to counteract this sensation and so concluded that the popular belief that attributed revivifying properties to the lake was completely unfounded.[31]

With this self-experimentation practice, Nollet offered himself as a testimonial of the power of experimental philosophy. He demonstrated that he literally bet his life that the knowledge deriving from the instruments of experimental philosophy could be trusted, just like early supporters of smallpox inoculation defended the practice by publicly subjecting themselves or their children to it.[32] Readers knew that animals subjected to experiments with the air pump recovered all vital functions when they were let out to breathe. Nollet decided that although the nature of the vapor remained unknown, it was obviously composed of a substance that impeded breathing.[33] His conclusion was so persuasive that it became a model for other enlightened naturalists. Years later, La Condamine and Lalande, his colleagues at the Paris Academy of Sciences, decided to replicate the experiments he had carried out during their own visit to the Cave of the Dog. They repeated them successfully and then subjected their own bodies to the mysterious vapor, reaching the same conclusions as Nollet.[34]

The possibility of contributing new explanations for the celebrated Italian marvels of nature was as tantalizing for a naturalist like Nollet as the hunt for antiquities or precious artworks for grand tourists. While grand tourists sought material souvenirs of their journeys, ambitious naturalists who obtained new knowledge during their visit relied on the presentation of results on the printed page to showcase their efforts. In travel accounts just as in experimental ones, the literary technology of virtual witnessing was crucial to the acquisition of authority.[35] Readers of Nollet, Lalande, or Brydone ideally traveled along with the authors and saw celebrated places in a new light. No longer sites where entertaining demonstrations reinforced the notion that they were natural marvels, naturalistic destinations became places where the instruments of experimental philosophy produced enlightenment. Thermometers, barometers, and other experimental apparatus not only demonstrated that nature was uniform even when it appeared strange or unusual but even more importantly confirmed that the experiments conducted in the artificial space of an experiment room—whether a salon or a lecture hall—revealed truths about parts of the natural world that audiences had not yet visited and perhaps never would.

In this journey of enlightenment, however, not everything was worth the philosopher's attention. Some marvelous accounts were blatantly false and could safely be ignored. During a trip in the outskirts of Rome, Nollet visited Monte Testaccio, a mound of broken pottery, tiles, bricks and pots from antiquity in which caves were dug that tavern keepers used to store their wine barrels. According to the Romans, the temperature inside these caves was extraordinarily low and the wine preserved there had the remarkable property of remaining cool longer than if chilled with ice. Nollet remarked that his thermometer revealed indeed that the temperature inside the cave was twenty degree lower than outside, but the idea that the wine could preserve its coldness longer than if chilled with ice was a "prejudice" that did not warrant "being battled with experiments."[36]

Nollet's fieldwork in the kingdom of Naples expanded the epistemological space of the laboratory. It crucially erased the boundary between the natural and the artificial, demonstrating that the conclusions reached in experimental settings could be applied to the world of nature. In his public presentation of his battle against the love of the marvelous, Nollet articulated for the French public the essential point that knowledge of experimental physics empowered them to spot potential frauds. Belief in natural marvels originated from ignorance or deception, whereas experimental physics provided instruments to sweep them away.

Miracles and Deception

Naples held a unique place in the eyes of foreign visitors not just because of its unique natural formations but also its people. Descriptions of local individuals had been commonplace in travelogues since the time of Pliny, and the Neapolitan *popolo*, the crowd, never failed to catch the tourists' attention. Representations of the Neapolitan crowd, which circulated widely once in print, played a central role in the exoticization of the city. Calaresu has vividly documented the travelers' fascination with the Neapolitan poor, the *lazzaroni*, which came to define in travel accounts a distinctive Neapolitan character. Travelers estimated that the poor amounted to a tenth of Naples's population at a time when the city was the third most populated in Europe. In the second half of the eighteenth century, guidebooks to Naples extended the stereotype of the *lazzaroni* as naturally idle, immoderate, and debauched to the entire "Neapolitan Nation."[37] Montesquieu's *The Spirit of the Laws*, which offered a naturalistic framework with which to account for local customs and temperament, fueled the association between the volcano and Neapolitans' character. Naples and its inhabitants were increasingly understood in terms of their similarity to Vesuvius: their idleness could readily turn to explosive eruptions without any intermediary stage.[38]

Climate theory was only one of the lenses through which travelers made sense of the peculiar customs they encountered in Naples. Tourists who arrived in the city before 1748 or those like Nollet, who did not think much of Montesquieu's theory, also remarked on the Neapolitans' loud character and what appeared in their eyes as an excessive religiosity. What struck them the most, though, was the crowd's susceptibility to deception and political manipulation. "A foreigner struggles to attend mass without distractions," Nollet remarked on his arrival in Naples. The local faithful gestured in ways that "elsewhere would be taken as eccentricities": both men and women hit "their chests strongly with their fists, [kissed] the ground constantly, [sighed and spoke] out loudly, in any tone whatsoever, to the good God and the Blessed Virgin Mary."[39]

This expressive devotion reached its peak on occasion of the liquefaction of the blood of St. Januarius, the saint patron of Naples. This extraordinary event occurred at set dates every year, and so it gave travelers the rare opportunity to plan the witnessing of a miracle. Kept in a glass phial inside a silver reliquary, the saint's blood was normally in a solid state. Three times a year and also during special circumstances such as calamities or visits of illustrious people, the blood inside the phial liquefied. This miraculous occurrence signaled the saint's renewed protection of the city, while failed liquefaction was widely understood as a warning message about forthcoming disasters such as famine, pestilence, volcanic eruptions, or social upheavals. The wait was therefore a particularly dramatic moment for the population. The miracle happened after several masses that the devotees participated in with distinctive, yet partially staged, intensity. Professional female weepers, the *prefiche*, enhanced the solemnity of the wait, as in ancient Mediterranean funeral traditions. Their role was generally lost on foreign travelers.[40]

The performative aspect of the Neapolitans' religiosity as manifested during the wait captured travelers' attention even more than the miracle itself and became a literary motif through which authors presented the special character of the Neapolitans. Protestant travelers did not hesitate to label the miracle a "spiritual farse," while French tourists were more cautious in expressing their views until the later part of the century, partly owing to the politically charged discussions on the *convulsionnaries* of Saint-Médard—a group of Jansenists who claimed that the tomb of the Jansenist deacon François de Pâris at the cemetery of Saint-Médard was a site of miraculous cures.[41] Montesquieu suspended his judgment on whether the liquefaction was indeed a miracle and decided not to divulge his impressions, but he privately commented on the impressionability of the Neapolitan faithful, particularly the *lazzaroni*, who feared the failed miracle more than anyone else.[42] Charles de Brosses, on the other hand, regarded the miracle "a nice piece of chemistry" and commented on the laziness of the Neapolitan people who had nothing but miracles, to occupy themselves with.[43] Similarly, Nollet chose not to publish his observations on the miracle, but he took extensive notes that he most likely used in private conversations with the Parisian elites.

These private notes indicate the extent to which the stereotyping in travel accounts had turned the Neapolitan crowd into a touristic attraction. The *ars apodemica* manuals prescribed that travelers describe the people they saw, and tourists visiting Naples certainly felt compelled to record their impressions in their diaries. During the first mass before the liquefaction of the blood of St. Januarius, Nollet heard a chorus of disconcerting wails when the congregants saw that he blood was still solid: "All I managed to hear in the mass of noise was that it had been agreed that the vial the blood of Saint Januarius was hard and not liquid: *duro duro duro*."[44] When the blood failed to liquify, "the people wept and lamented as if they were in a shipwreck": from every corner prayers and supplications went up; the priest recited the *De profundis* "with a devotion so great that tears rolled down his cheeks," while the women facing the balustrade "were writhing about and screaming in a frightful way."[45]

The detailed descriptions of the devotees enmeshed in these intense emotions positioned the observer as an outsider whose detached observations authoritatively accounted for a collective deception orchestrated by the church hierarchy. Enlightened discussions on miracles centered on testimony more than the possibility of miracles themselves.[46] The public nature of the miracle of St. Januarius, then, offered opportunities to expose abuses perpetrated by religious authorities. In Nollet's as in other travelers' accounts, the miracle of St. Januarius was presented as an obvious deception in which the old tools of natural magic were put in the service of political agendas. Observers familiar with chemistry saw the miracle as an unsurprising demonstration in which heat melted a solid substance. According to Nollet, the priest passed the relic to "all bystanders, placing it at the mouth and forehead of each person" and then put it on the altar with two candles at its sides. Finally, "by dint of kissing, handling and illuminating the relic with the candle . . . whatever was inside the inverted vial gradually oozed out like a paste that softens, . . . just as a hardened material normally does when it begins to melt in a container whose walls are heated." He labeled the entire ceremony a ritualized "deception" and a shameful abuse of popular credulity.[47]

The miracle of St Januarius was a powerful lens through which foreign travelers understood Naples and its inhabitants. For those who understood it as a collective deception, it introduced an element of skepticism that extended well beyond the stereotyping of the *lazzaroni*. Seen as complicit in this large-scale abuse, the Neapolitan elites—religious, aristocratic, or learned—appeared less deserving of credit than in other cities. In particular, the investment of learned aristocrats in enlightened natural philosophy appeared as a superficial display of patronage more than a real commitment to a project of improvement and education.

Nollet's comments on a number of his Neapolitan high-ranking acquaintances prove this point. In his diary he observed that the prince of Tarsia's taste for golden decorations made it impossible to read the titles of the books in his celebrated library. The prince of San Severo, meanwhile, with his claims to alchemical knowledge, did not seem worthy of membership in the Paris Academy of Sciences, which the prince so eagerly sought. Even Cardinal Spinelli aroused Nollet's suspicions. When the blood failed to liquefy on the first day, the cardinal somberly retired to his chambers, publicly claiming that he would be intensely praying to St. Januarius. Although he had made it clear that he would be available to no one, he readily made an exception for Nollet when he heard that the abbé brought a letter for him on behalf of some important personality in Rome. In his private journal, Nollet remarked that after reading the letter, the prelate looked so cheerful that one might justifiably suspect that "he was not in as much pain as he had told us." Or, he added in his diary, the cardinal "had reason to believe that the miracle would be more successful the next day."[48]

The ostentatious nature of Neapolitans' religiosity reminded visitors of superstitious rituals and made more of an impression than the events surrounding the liquefaction of St. Januarius's blood. "More attached than other Italians to exterior cult, and to all that is called devotional practice," the Neapolitans donated much money to have their churches decorated. Their religious fervor was a potential threat to public order and made them susceptible to collective manipulation. For example, Francesco Pepe, a Jesuit friar, had so much of a following that "it would be dangerous for the government to do anything that would displease this clergyman." The preacher took advantage of his extraordinary popularity to collect contributions to build an "obelisk in terrible taste."[49] What appeared in their eyes as a hazy boundary between devotion, superstition, and malfeasance encouraged foreign travelers to direct their attention to the people in Naples more than the people in any other Italian city. Nollet, who only rarely commented on the local customs, noted in his diary that the markets and streets of Naples bustled with "home nuns," women who wore a religious gown without taking any orders and worked as domestic servants. These "beguines" walked in small groups carrying a crucifix as if in a procession, "only to buy a piece of meat."[50]

The reporting on this unusual urban life contributed to othering Naples as a frontier of European civilization. Tourists absorbed the stereotypes while reading travel accounts in preparation for their trip and reinforced them with their own tales, in conversations or in print. Travel narratives circulated the idea of Italians as lovers of the marvelous, but in Naples the marvelous acquired a distinctive "exotic" flavor. Some travelers used climate theories to naturalize the uniqueness of the Neapolitan character through its association with Vesuvius.[51] Other naturalists, like Nollet, singled out Naples as the place where a deeply rooted love of the marvelous hindered the spread of the enlightened project of improvement through education.

Neapolitan Connections

If travel narratives exoticized Naples and the Neapolitans, the less public encounters documented in Nollet's diary indicate that he and the local learned community shared the culture of reciprocity that characterized the Republic of Letters. Nollet visited Naples in his public role as a member of the Paris Academy of Sciences and a secret consultant of the French Bureau of Commerce. On his arrival, he met the French ambassador in Naples, who explained that the French state had already received plenty of information about the local silk manufacture and discharged him of any intelligence-gathering duties. Nollet spent his time in Naples building connections with the local community of naturalists and savants. In this endeavor, he complied with the recommendations and requests of his colleagues in Paris. The secretary of the Academy of Sciences, for example, gave him a number of topics to discuss with the famous Neapolitan doctor Nicola Cirillo.[52] Unlike armchair travel, real-life visits to foreign places unfolded through personal connections with local individuals. Access to the local community through letters of introduction, visits to *conversazioni*, attendance at theater and music performances, and strolls in the company of local personalities were all essential to the success of the visit. Particularly crucial for naturalists like Nollet, locals provided access to fieldwork research and offered connections that would remain in place even after the journey.

The interest in forming connections was reciprocal. The visit of a foreign celebrity such as Nollet was a compliment for local individuals. Neapolitan nobles and savants who shared an interest in natural knowledge competed for Nollet's attention. Local aristocrats who sponsored natural philosophy hoped that their encounter with the distinguished visitor might give them greater visibility in the Republic of Letters and, as a consequence, bestow more relevance on them at the local level. Raimondo di Sangro, the prince of San Severo, for instance, repeatedly invited Nollet to his palace, openly manifesting the desire—which he never abandoned—to be elected a foreign member of the Paris Academy of Sciences. Another local aristocrat that Nollet met was the prince of Scalea, Francesco Maria Spinelli, who decades earlier had supported the establishment of a scientific academy directed by the archbishop of Taranto, Celestino Galiani. Spinelli was the author of several treaties on Cartesian philosophy and owned a collection of scientific instruments. Nollet was impressed by the prince's enthusiasm for natural philosophy, and after an evening in his company, he noted in his diary that the prince, unlike most aristocrats who attended experimental demonstrations, had performed the "experiments himself."[53]

Nollet also socialized with the group of intellectuals who met at the home of Faustina Pignatelli, the princess of Colubrano, a well-known learned woman who was an honorary member of the Bologna Academy of Sciences and the protagonist of an imaginary conversation on the controversy of "living forces" that the Academy's secretary, Francesco Maria Zanotti, published in 1752.[54] Parisian mathematicians too were familiar with Pignatelli, who was a correspondent of De Mairan. Pignatelli's *conversazione* in Naples was attended by a circle of philosophers and mathematicians who had previously overseen the princess's scientific training and who, thanks to their commitment to disseminating Newtonian philosophy, had received words of praise from Voltaire. Nollet enjoyed his conversations with Francesco Serrao, a former member of the Neapolitan Academy of Sciences, who "had written so well on Vesuvius and tarantulas."[55]

The most enduring relationship that Nollet formed in Naples was with Mariangela Ardinghelli, a twenty-one-year-old learned woman whose reputation in the Republic of Letters was just emerging. Ardinghelli was the only daughter of a patrician family of Tuscan origin, and had advanced in her scientific learning thanks to a coterie of ambitious local academicians that gathered in the library of the prince of Tarsia, where she participated in experiments. At the time of Nollet's visit, Ardinghelli had just completed the Italian translation of Stephen Hales's *Haemastaticks* (1733), a key text that applied Newtonian philosophy to human physiology. Her work was more than a simple translation, as she included various notes that showcased her own knowledge and understanding. During his stay in Naples, Nollet often ended his evenings at Ardinghelli's *conversazione* and developed with her a relationship that lasted a lifetime. Their correspondence demonstrates that she acted as an informal foreign correspondent for the Paris Academy of Sciences, an institution that did not admit women. Over the course of two decades, Nollet read numerous excerpts from Ardinghelli's letters that he had translated into French to the Academy in Paris. In 1760, he dedicated the first letter in his *Lettres sur l'électricité* to her and, in turn, she translated Nollet's book into Italian. He admired her so much that he commissioned a medallion with her portrait, which he hung in his physics cabinet.[56]

The Neapolitan electricians who gathered in the palazzo of the prince of Tarsia eagerly awaited Nollet's arrival. Palazzo Tarsia was one of the most active circles for scientific sociability in Naples. There, Della Torre performed electrical experiments for local and foreign aristocrats, with the collaboration of Bammacaro and Ardinghelli. Although Nollet had criticized Bammacaro's electrical theories published in *Tentamen de vi electrica ejusque phaenomeni* (An examination of the electric force and its phenomena), the two developed an amicable relationship during the sumptuous reception organized in Nollet's honor by the prince of Tarsia. The prince also received Nollet privately and commissioned a fair number of instruments from him. In private, Nollet snubbed the prince's "little collection of instruments," but he was pleased to take the commission.[57]

The Reciprocity of Stereotypes

No philosophical duel ensued between Nollet and his Neapolitan counterparts, yet beyond the façade of mutual interest and academic exchange, the encounters generated irritated responses. The king of Naples's prime minister, Bernardo Tanucci, who had not met Nollet, reported to a friend in Tuscany that because of the abbé's arrogance and lack of Italian, "the Neapolitans despise him." In Tanucci's interpretation, arrogance and ignorance of the Italian language were not unique to Nollet: they were typically French. Just as Nollet and other travelers stereotyped the Italians as lovers of the marvelous and the Neapolitans as idle and explosive like Vesuvius, Tanucci employed stereotypes about the French that circulated in his milieu. The target of his attack was France's linguistic imperialism, coupled with the French monarchy's self-presentation as the new Rome: "However big the French have become by means of gunshots and tricks, they cannot compare to the ancient Romans, . . . who did not go to Asia without having Greek," he commented.[58] Tanucci's remarks were quite typical. On his visit to Paris, the German philosopher Johann Gottfried Herder similarly complained about the linguistic regime that the French had imposed on the Republic of Letters.[59] Pivati employed similarly nationalistic rhetoric when he warned Veratti about Nollet's preconceived ideas about the medicated tubes: he described Nollet as full of prejudices that were typical "of his nation."[60]

These reverse stereotypes increased with the number of publications about Naples and its surroundings. Political reformers were dismayed at the craze for naturalistic excursions, which made foreign travelers blind to the intellectual ferments ongoing in the city. The economist Ferdinando Galiani bitterly commented on the tourists who "do nothing but see a few bricks and bits of marble at Pozzuoli and Portici, a few smoldering rocks at the Solfatara and the Vesuvius, a day at San Martino, a night at the theatre and in eight days they have dispatched with everything."[61] Meanwhile, local naturalists, who had published mostly in Italian to obtain the favors of local patrons, felt threatened by the growing popularity of foreign works on the natural phenomena around Naples. They responded in print to their colleagues abroad, underscoring inaccuracies and errors in order to reaffirm their privileged position vis-à-vis local marvels. In his *Storia e fenomeni del Vesuvio*, Della Torre pointed to several mistakes in Nollet's report about Vesuvius, emphasizing that they no doubt owed to the fact that the abbé's visit was too short for him to become deeply acquainted with the volcano. Della Torre believed that the geophysical phenomena around Naples were unique to the place and should be studied locally. His approach was incompatible with Nollet's idea that instruments like the fire engine or the aeolipile could be taken as models of volcanic phenomena across the globe. Nollet intended to correct data produced by local experimenters, but Della Torre underscored that the abbé had advanced an unsound theory about the formation of volcanos based on a few hasty measurements.[62]

Tourists' travel accounts did not always please Italian readers, who doubted the ability of foreigners to produce a narrative about their journey to Italy that was "judicious, historical, impartial and stripped of the innkeepers' fables, blindly accepted."[63] These complaints only increased over the course of the century, culminating in Pietro Napoli-Signorelli's sharp criticism: "Seeing that travel writers always repeat the same things by copying their predecessors without fact-checking, one can legitimately and plausibly suspect that they finished their books before seeing Italy, and that they then travel just to confirm, if even possible, what insults it, without regard to the evident truth that flies in the face of what they recorded before crossing the Alps."[64] This trenchant assessment certainly applied to Nollet's published account. His presumed battle against the love of the marvelous was a scripted tale planned well before his journey. Similarly, his local counterparts had their biases against foreigners, for whom they constructed prepackaged experiences that built on the imagined reality of Italy as a land of marvels. Reciprocity was a founding principle of the Republic of Letters, and in this two-way relationship, all parties relied on stereotyping.[65]

Conclusion

The published version of Nollet's journey through Italy was a largely fabricated story that resonated with cultural stereotypes created by travel narratives: the Italians' love of the marvelous, which tourists knew they had to guard themselves against, also tainted experimental philosophy. Yet beyond the fabricated version he offered the reading public, Nollet's journey through the Italian states was a successful business trip. Like the merchants who traveled in search of rare pieces to resell to collectors when they returned home, Nollet collected news and gathered information that he exchanged for social and economic returns. In addition to the recognition he earned for the reports he delivered to the French Bureau of Commerce and the Paris Academy of Sciences, Nollet received several commissions from Italian prelates, aristocrats, and savants. The cardinal Domenico Passionei in Rome ordered a microscope, a barometer, and two different types of thermometers, a "signor Gueraldi" ordered two mercury thermometers, and the marquis Joseph Philippe d'Oncieux requested a pair of globes and a microscope, for which, he specified, he expected to pay a lot. The prince of Tarsia gave Nollet carte blanche to add instruments to his collection, and the king of Piedmont, who had already commissioned an entire physics cabinet from him, entrusted him with checking the famous Pestalozzi collection to see if there were some pieces there that would be appropriate for the royal cabinet of curiosities. As a sign of gratitude for all of Nollet's services, Carlo Emanuele III granted him the high honor of the cross of St. Maurice.[1]

Nollet acted also as an efficient envoy of the Paris Academy of Sciences. His journey brought three new foreign correspondents to the Academy: Francesco Garro, Francesco Maria Zanotti, and Giovanni Maria Della Torre. Nollet also added François Jacquier, Carlo Alfonso Guadagni, Giambattista Beccaria, and Ignazio Sommis to his own personal correspondence network, all of whom offered to write to him every two months to keep him abreast of the state of the sciences in their cities.[2] Most relevant among these new personal connections was Mariangela Ardinghelli, who acted for two decades as an informal foreign correspondent from Naples for the Paris Academy of Sciences. Nollet also exchanged a few letters with Laura Bassi and helped to consolidate institutional relations with the Bologna Academy of Sciences, which committed to subscribing to the Paris Academy of Sciences' periodical publication.[3]

In the published version that circulated widely among the eighteenth-century reading public, Nollet's journey was motivated by a battle in the name of truth against the love of the marvelous. This book has revealed that both the episodes that sparked the controversy and the philosophical duel were fabrications that Pivati and Nollet purposefully constructed with the goal of boosting their reputations. Although they were rivals, they both shared communication strategies rooted in the power of the printed page to create fabricated realities that many readers took at face value. Pivati hoped that his electric cures would work as an effective advertisement campaign for his expensive encyclopedic dictionary in a competitive book market. He published reports of experiments that he had carelessly performed—and never repeated—and concealed the information that the relief that his medicated tubes seemed to offer was only temporary, if at all real. Pivati drew on the strategies of charlatans whose writings he reviewed in his official position as the superintended to the book trade, embellishing his accounts and even outright lying. The fabricated reality he created on the printed page struck a chord with the local reading audience. His presumed invention attributed to Italy priority in the worthwhile effort of turning electricity from an entertaining spectacle into a useful science. It also earned him the support of the Bologna Academy of Sciences, which was crucial to the reception of the medicated tubes among the electricians that operated in other institutions, in Italy, and in the rest of Europe.

Nollet, in turn, constructed a version of his journey through Italy in which he represented himself as an expert arbiter in the emerging field of electricity—a version that effectively served as a cover for the secret mission he carried out on behalf of the French state. The story of the sham tubes fed the experimental philosophers' anxiety about the persistence of marvelous accounts in the learned world. In declaring "the marvelous is not for us," Diderot and d'Alembert implicitly pointed to its persistence and pervasiveness.[4] In the natural world, as in the realm of the healing arts, many baffling and surprising phenomena seemed to undermine the fundamental premise upon which experimental philosophy was based: the idea of a natural world that followed the same rules independently of place or time.

The philosophical duel staged by Nollet, as well as his reports on the natural phenomena he observed in the surroundings of Naples, had the reassuring virtue of confirming the key principle of the universality of the laws of nature. This constituted the epistemological foundation of the educational programs experimental philosophers like Nollet offered to learned elites. Demonstrations of experiments that audiences witnessed in Nollet's home in Paris offered tools for understanding phenomena that took place in other sites, for example, at sea, or in distant colonies, or in the Campi Flegrei. The experiments carried out in Paris and London, which disproved the claim that the medicated tubes had curative value, also demonstrated that there could be nothing special about Italy that made the medicated tubes work only there—except an excessive love of the marvelous.

In the fabricated version of Nollet's journey, marvels became synonymous with deception and self-deception. Those who deceived themselves by not engaging with the methods of experimental philosophy were branded as lovers of the marvelous. Defending the ideal of a natural world whose operations followed universal, eternal laws, required considerable efforts in a public arena in which news of earthquakes, volcanic eruptions, prodigious cures, monstrous births, and other unusual phenomena continued to captivate the readers' imaginations. Natural philosophers eagerly defined the proper methods for establishing matters of fact. In their efforts, they had to compete for space and attention with other historical actors who actively spread false information. Famous hoaxes, from Mary Tofts (the woman who allegedly gave birth to rabbits) to Johann Batholomeus Adam Beringer's lying stones, were warning tales for the learned, who realized they too could fall prey to deceit.[5] Who to believe and what—the problem of trust, which has been central in studies on the emergence of experimental philosophy in the seventeenth century—had to do with social status just as with the long-lasting tradition of erudite forgeries, fake news, and hoaxes.[6] In this context, Nollet's fabricated exposé acquired a significance that went well beyond the specific case of the medicated tubes. It presented the experimental philosopher not simply as a disinterested gentleman but above all as an open-minded expert who could reliably detect the ever-looming threats of deception and self-deception. Because experimental philosophy could be learned virtually by everyone, the experimental philosopher was also a means to the crucial end of achieving enlightenment.

This book has taken the creation of these fabricated realities as the subject of historical analysis. It has shown that the science of electricity, as the new science of the eighteenth century, offered opportunities for newcomers in the field of experimental philosophy to make a name for themselves. In the absence of experimental protocols to apply, ancient authorities to confront, or standard instruments to master, self-styled electricians and medical electricians advanced extraordinary claims that often resulted in tangible returns. To audiences that thought of themselves as a tribunal of public opinion, controversies mattered as much as, if not more than, the establishment of new matters of fact. Controversial experiments prompted some readers to become performers and boosted interest in the new electrical science. In these early stages, when electrical experiments were relatively easy to conduct and instruments not difficult to build, electricity created also new forms of sociability. The casino of Querini in Venice, the villa of the count Gazola and his wife the countess Guarienti in Verona, the library of the prince of Tarsia in Naples, the Accademia dei Vari in Bologna, and the house of Mariotti's colleague in Perugia all became spaces in which intellectual ambitions, experimental practice, and social engagements converged. These early ferments constituted the foundations on which the more celebrated works of Italian electricians such as Giambattista Beccaria, Luigi Galvani, and Alessandro Volta took off only a few years later.

Nollet's journey—both its secret and its official version—reveals that the paradigm of the Grand Tour can act as an intellectual straitjacket, obscuring the various other features that made eighteenth-century Italy a travel destination. Naturalists and savants crossed the Alps to interact with their colleagues in the Italian states and to study unique natural phenomena. They conducted fieldwork and gathered information—both secretly and openly—on Italian manufactures, crafts, mines, crops, and people. The numerous publications that ensued from such activities testify to a widespread learned interest in Italian scientific culture that, with notable exceptions, is still not reflected in the Anglophone scholarly literature on eighteenth-century science and medicine. My intention, however, has not been so much to call attention to eighteenth-century Italian science as a fertile topic of scholarly investigation as to highlight the important role that travel and travel accounts played in the creation and circulation of stereotypes within Europe. Travel accounts by naturalists such as Nollet presented the locals' understanding of natural phenomena in terms of an outdated love of the marvelous, and in turn implicitly defined their own approach as modern and enlightened. In doing so, they fabricated stories in which the complex negotiations and human interactions that characterized experimental philosophy were invariably absent or irrelevant.

These kinds of accounts built consensus on the idea that science needed a specific kind of history: one that did not account for real episodes, especially, as Joseph Priestley emphasized, when they revealed that human passions could get in the way of progress. What mattered in historical accounts about the sciences was providing a lesson for the future, even at the cost of accuracy. These fabricated realities reified a notion of scientific development as a linear progress toward a self-evident truth that readers were called on to simply trust rather than challenge or interrogate. This "scientific exceptionalism"—the idea that science differs from other forms of human learning and needs a special kind of history—is still very pervasive. In our times of alternate facts, conspiracy theories, and fake news, these eighteenth-century episodes suggest that more transparency about the practices through which scientific results are established—together perhaps with broader education in the history and sociology of scientific knowledge—might in fact be a more effective strategy for addressing planetary emergencies than an acritical defense of scientific truth.

Abbreviations

AASB	Archivio dell'Accademia delle scienze, Bologna
AASP	Archives de l'Académie des sciences, Paris
ANP	Archives nationales, Paris
AST	Archivio di Stato, Turin
ASUT	Archivio storico dell'Università, Turin
BCAB	Biblioteca comunale dell'Archiginnasio, Bologna
BGR	Biblioteca civica Gambalunga, Rimini
BMN	Bibliothèque municipale, Nîmes
BMS	Bibliothèque municipale de Soisson
BNM	Biblioteca nazionale Marciana, Venice
BPUG	Bibliothèque publique et universitaire, Geneva
BRT	Biblioteca reale, Turin
PV	Procès Verbaux, AASP
RSL	Royal Society Library, London

Introduction

1. Pivati, *Della elettricità medica lettera del chiarissimo Signore Gio. Francesco Pivati.*

2. Nollet, "Expériences et observations en différens endroits d'Italie."

3. Beretta, Clericuzio, and Principe, *The Accademia del Cimento and Its European Context*; Findlen, Wassyng Roworth, and Sama, *Italy's Eighteenth Century*; Daston and Park, *Wonders and the Order of Nature.*

4. BMS, MS 150, Nollet, "Journal du voyage de Piedmont et d'Italie en 1749." A transcription of the diary can be found at https://www.press.jhu.edu/books/title/12934/land-marvels#book _resources.

5. Bertucci, "Enlightened Secrets."

6. Jouhaud, *Mazarinades*; de Vivo, *Information and Communication in Venice*; Darnton, "An Early Information Society."

7. Shapin and Schaffer, *Leviathan and the Air-Pump*; Bertucci, *Viaggio nel paese delle meraviglie.*

8. Johns, *The Nature of the Book*, 457.

9. Grafton, *Forgers and Critics*; Stephens, Havens, and Gomez, *Literary Forgery in Early Modern Europe*; Beretta and Conforti, *Fakes!?*; Lynch, *Deception and Detection in Eighteenth-Century Britain*; Donato, *The Life and Legend of Catterina Vizzani.* See also Blair, Duguid, Goeing, and Grafton, *Information.*

10. Darnton, "The True History of Fake News"; Soll, "The Long and Brutal History of Fake News"; Lynch, *Deception and Detection in Eighteenth-Century Britain*; Butterworth, *Poisoned Words*. See also the relevant articles in Blair, Duguid, Goeing, and Grafton, *Information*.

11. Infelise, "Criminali e cronaca nera negli strumenti pubblici di informazione tra '600 e '700"; Lever, *Canards sanglants*; Natale, *Gli specchi della paura*; Caracciolo, "Notizie false e pratiche editoriali negli avvisi a stampa di antico regime."

12. Alder, "History's Greatest Forger."

13. Eisenstein, *The Printing Press as an Agent of Change*; Johns, *The Nature of the Book*.

14. Darnton, "Policing Writers in Paris circa 1750"; Darnton, "An Early Information Society"; Darnton, *Forbidden Best-Sellers of Pre-Revolutionary France*; Berengo, *Giornali veneziani del Settecento*; Infelise, *L'editoria veneziana nel '700*.

15. Oreskes and Conway, *Merchants of Doubt*; Proctor and Schiebinger, *Agnotology*; Ceccarelli, "Manufactured Scientific Controversy"; Bar-Ilan and Halevi, "Retracted Articles."

16. Füssel, "'The Charlatanry of the Learned.'"

17. Nollet, "Extract of a Letter from the Abbé Nollet, F. R. S., etc. to Charles, Duke of Richmond, F. R. S., Accompanying an Examination of Certain Phaenomena in Electricity, Published in Italy," 397.

18. Daston and Park, *Wonders and the Order of Nature*, chap. 7.

19. Sweet, Verhoeven, and Goldsmith, *Beyond the Grand Tour*; Findlen, Roworth, and Sama, *Italy's Eighteenth Century*. The literature on the Grand Tour is vast. The bibliographies in *Beyond the Grand Tour* and *Italy's Eighteenth Century* are great starting points.

20. Nollet's accounts of his journey to Italy, which he read at the Paris Academy of Sciences, are published in Nollet, "Expériences et observations en différent endroits d'Italie," and Nollet, "Suite des expériences et observations en différent endroits d'Italie."

21. Brilli, *Quando viaggiare era un'arte*; de Seta, "Grand Tour"; Maçzak, *Viaggi e viaggiatori nell'Europa moderna*; Brilli, *Il grande racconto del viaggio in Italia*; Bertrand, *Le Grand Tour revisité*; Sénelier, *Voyageurs français en Italie du Moyen Age à nos jours*; Brizay, *Touristes du Grand Siècle*; Chapron, "Voyageurs et bibliothèques dans l'Italie du XVIIIe siècle des *Mirabilia* au débat sur l'utilité publique."

22. BMS, MS 150, Nollet, "Journal du voyage de Piedmont et d'Italie en 1749," ff. 105–6, 46. See also Nollet, "Suite des expériences et observations en différent endroits d'Italie," 57ff.

23. Bianchini, "Extrait des observations faites au mois de Décembre 1705 par M. Bianchini, sur des feux qui se voient sur une des montagnes de l'Apennin."

24. Cavazza, "'Philocentria' e Pietramala."

25. Misson, *Nouveau voyage d'Italie de Monsieur Misson*.

26. Brosses and Babou, *Lettres familières écrites d'Italie à quelques amis en 1739 et 1740*, 268.

27. Stagl, "Ars Apodemica and Socio-Cultural Research."

28. Di Mitri, *Storia biomedica del tarantismo nel XVIII secolo*.

29. Casillo, *The Empire of Stereotypes*.

30. Misson, *Nouveau voyage d'Italie de Monsieur Misson*.

31. Brosses and Babou, *Lettres familières écrites d'Italie à quelques amis en 1739 et 1740*.

32. Macchia and Colesanti, *Montesquieu*, 118.

33. Nollet, "Suite des expériences et observations en différent endroits d'Italie," 483.

34. Brosses and Babou, *Lettres familières*, 235.

35. Nollet, "Suite des expériences et observations en différent endroits d'Italie," 483.

36. Baretti, *An Account of the Manners and Customs of Italy*, 1:3.

37. Joau-Pau Rubiés, "Comparing Cultures in the Early Modern World"; Rubiés and Ollé, "The Comparative History of a Genre"; Rubiés, "Nature and Customs in Late Medieval Ethnography"; Davies, *Renaissance Ethnography and the Invention of the Human*.

38. Calaresu, "From the Street to Stereotype."

39. Jones, *Industrial Enlightenment*; Mokyr, *The Enlightened Economy*; McClellan and Regourd, *The Colonial Machine*; Beaurepaire and Pourchasse, *Les circulations internationales en Europe*. For a global perspective, see Schaffer, Roberts, Raj, Delbourgo, and Sibum, *The Brokered World*, Roche, *Les circulations dans l'Europe moderne*, Roche, *Humeurs vagabondes*, and Withers, *Placing the Enlightenment*. On travel and scientific culture, see Terrall, *The Man Who Flattened the Earth*, Bourget, Licoppe, and Sibum, *Instruments, Travel, and Science*, and Hesse and Sahlins, "Mobility in French History."

40. Sterne, *A Sentimental Journey*, 12.

41. *Dictionnaire de l'Académie Françoise*, 1:524

42. On Pidansat de Mairobert, see Darnton, *Forbidden Best-Sellers of Pre-Revolutionary France*.

43. Etymological dictionaries published in the late eighteenth and early nineteenth centuries traced the origin of the verb "to spy" to the German "spæhen" and the Latin "aspicere," both of which mean "to observe" (Ménage, *Dictionnaire étymologique*, 1:541; de Roquefort, *Dictionnaire étymologique de la langue françoise*, 1:278).

44. On the state of secrecy enforced by Colbert, see Soll, *The Information Master*.

45. Montesquieu, *Oeuvres de Montesquieu*, 173. See also Bély, *Espions et ambassadeurs au temps de Louis XIV*.

46. Hilaire-Pérez, "Diderot's Views on Artists' and Inventors' Rights."

47. Diderot and d'Alembert, *Encyclopédie*, 5:971.

48. Anon., *Traité des ambassades et des ambassadeurs*, 156–57.

49. Goudar, *L'espion francois à Londres*, 1:1.

50. Goudar, *L'espion françois à Londres*, 1:1–8. On Goudar see Hauc, *Ange Goudar*.

51. Quoted in Hauc, *Ange Goudar*, 25.

52. Harris, *Industrial Espionage and Technology Transfer*. For criticism of Harris's approach in the context of industrial development, see Bret, Gouzevitch, and Hilaire-Pérez, *Les techniques et la technologie*, Hilaire-Pérez, "Les échanges techniques entre la France et l'Angleterre au XVIIIe siècle," and Hilaire-Pérez and Verna, "Dissemination of Technical Knowledge in the Middle Ages and the Early Modern Era."

53. Chicco, *La seta in Piemonte*; Bertucci, "Spinners' Hands, Imperial Minds."

54. On the public culture of science, the foundational works are Stewart, *The Rise of Public Science*, Golinski, *Science as Public Culture*, and Schaffer, "Natural Philosophy and Public Spectacle in the Eighteenth Century." Historians such as Margaret Jacob and Larry Stewart (*Practical Matter*) as well as Joel Mokyr (*The Enlightened Economy*) have linked the openness of scientific knowledge to the British Industrial Revolution.

55. Jürgen Habermas's notion of the public sphere has elicited a flurry of historiographical discussions among scholars of the eighteenth century, who have repeatedly emphasized the public dimension of Enlightenment political and cultural debates. As Thomas Broman points out in "The Habermasian Public Sphere," however, the emphasis on the public in the history of science was not directly informed by the notion of the public sphere but rather by the sociology of scientific knowledge. See Habermas, *The Structural Transformation of the Public Sphere*. It is impossible to list here all the works that have been inspired by Habermas's concept of the public sphere, but see at least Calhoun, *Habermas and the Public Sphere*, and Melton, *The Rise of the Public in Enlightenment Europe*.

56. Margocsy and Vermeir, "State of Secrecy."

57. On the problematic distinction between openness and secrecy up to the sixteenth century, see Long, *Openness, Secrecy, Authorship*. Jacob Soll (*The Information Master*) has urged historians to pay more attention to the relationship between the public sphere and state secrecy.

58. Harris, *Industrial Espionage and Technology Transfer*, 527.

59. Long, *Openness, Secrecy, Authorship*.

60. Leong and Rankin, *Secrets and Knowledge in Medicine and Science*; Eamon, *Science and the Secrets of Nature*; Principe, *The Secrets of Alchemy*; Kavey, *Books of Secrets*.

61. Edward Gibbon, *Viaggio in Italia*.

Chapter 1 · Silk and Secrets

1. ANP, F12/1453, "Observations sur le filage des soies," Gaudin to Trudaine and Jouy, June 30, 1749.

2. See entry 4761 in du Sommerard, *Catalogue et description des objets d'art de l'antiquité du moyen âge et de la Renaissance*, 374. On Nollet's life, see de Fouchy, "Eloge de M. L'Abbé Nollet," Torlais, *Un physicien au siècle des Lumières*, Pyenson and Gauvin, *The Art of Teaching Physics*, and Piot, *Jean-Antoine Nollet*.

3. Bertucci, *Artisanal Enlightenment*; Bertucci and Courcelle, "Artisanal Knowledge, Expertise, and Patronage in Early Eighteenth-Century Paris."

4. A photo reproduction of one of Nollet's terrestrial globe, which shows the cartouche, is available at https://gallica.bnf.fr/ark:/12148/btv1b55008754z. On Nollet's globes, see Ronfort, "Science and Luxury."

5. Daumas, *Scientific Instruments of the Seventeenth and Eighteenth Centuries*.

6. Gauvin, "The Instrument That Never Was"; Terrall, *Catching Nature in the Act*.

7. de Fouchy, "Eloge de M. L'Abbé Nollet." The volume on glassmaking was never published and there is no evidence that Nollet ever started this work, but he is listed as the potential author in AASP, PV 77bis (1758), ff. 597–99. It would be worth investigating whether Nollet may have been involved in the essay on enameling, glassmaking, and glazing in the *Encyclopédie*, attributed to Diderot.

8. Stafford, *Artful Science*; Sutton, *Science for a Polite Society*; Bensaude-Vincent and Blondel, *Science and Spectacle in the European Enlightenment*; Stewart, *The Rise of Public Science*; Schaffer, "Natural Philosophy and Public Spectacle in the Eighteenth Century"; Schaffer, "Machine Philosophy"; de Clercq, *At the Sign of the Oriental Lamp*.

9. Nollet, *Programme*, xi.

10. Voltaire, *Correspondence*, 1:1058–59.

11. Voltaire, *Correspondence*, 1:1094. On Voltaire's cabinet, see Gauvin, "Le cabinet de physique du château de Cirey et la philosophie naturelle de Mme du Châtelet et de Voltaire."

12. Nollet, *Leçons de physique expérimentale*, xxii.

13. Nollet, *Leçons de physique expérimentale*, xxiv.

14. Nollet, *Programme*, xxxi, xxxii, xxxvi.

15. Algarotti, *Opere*, 16:16.

16. BRT, Varia 299, "Idée abregée des etudes faites par S.A.R. le duc Victor Amedée redigée l'an 1747." Nollet's arrival at court is documented in BRT, Storia Patria 932, Orioles, cav., "Giornale di quanto avvenne alla corte in Torino, dal 1714 al 1748," f. 98. See also Benguigui *Théories électriques du XVIIIe siècle*, 89.

17. Benguigui, *Théories électriques du XVIIIe siècle*, 87.

18. Ferrone, *La nuova Atlantide e i lumi*.

19. AST, Patenti Controllo Finanze, reg. 9, f. 174, "Giubilazione di Francesco Garro, professore di fisica sperimentale"; Archivio Storico dell'Università, Turin, Mandati di pagamento, 12 C1, ff. 119 sgg., AST, art. 217, "Conto camerale della tesoreria della casa di S. M. per l'anno 1741," anno 1741, n. 46.

20. Quignon, *L'abbé Nollet Physicien*, 66.

21. See the manuscript "Catalogo del gabinetto di fisica" at the Museo di Fisica dell'Università in Turin.

22. Anon., "Gabinetto di Fisica," 532; Roero, "Il Gabinetto di Fisica nel Settecento."

23. Roche, *Le siècle des Lumières en province.*

24. Torlais, *Un physicien au siècle des Lumières.*

25. Luynes, *Mémoires du duc de Luynes sur la cour de Louis XV*, 5:453.

26. ANP, O1 195, Décisions du roi-dépenses, ff. 36v e 129.

27. ANP, O1 90, Secrétariat de la maison du roi, f. 80; Daumas, *Scientific Instruments of the Seventeenth and Eighteenth Centuries*, 99.

28. Hahn, *The Anatomy of a Scientific Institution*; Shank, *Before Voltaire.*

29. AASP, Dossier Réaumur, "Réflexion sur l'utilité dont l'Académie des Sciences pourrait être au Royaume, si le Royaume lui donnait les secours dont elle a besoin." The document is not dated.

30. BPUG, Correspondance Jallabert, SHAG 242, f. 95, Nollet to Jallabert, January 20, 1740. The correspondence is also published in Benguigui, *Théories électriques du XVIIIe siècle.*

31. ANP, F12/1432A.

32. On silk in Lyon, see Doyon and Liaigre, *Jacques Vaucanson*, Ballot, *L'introduction du machinisme dans l'industrie française*, Sabel and Zeitlin, "Fashion as Flexible Production," and Hilaire-Pérez, "Cultures techniques et pratiques de l'échange, entre Lyon et le Levant."

33. BMS, MS 150, Nollet, "Journal du voyage de Piemont et d'Italie en 1749," f. 1.

34. On silk manufacture in Piedmont, see Chicco, *La seta in Piemonte.* On the quality of Piedmont's organzine, see Poni, "Standards, Trust, and Civil Discourse." On the long-term history of silk manufacture in Italy, see Molà, *The Silk Industry of Renaissance Venice*, and Molà, Mueller, and Zanier, *La seta in Italia dal Medioevo al Seicento.*

35. Davini, "Bengali Raw Silk, the East India Company and the European Global Market"; Hutková, "Technology Transfers and Organization"; Chicco, *La seta in Piemonte*; Bertucci, "Spinners' Hands, Imperial Minds."

36. Long, *Openness, Secrecy, Authorship*; Smith, "What Is a Secret?"; Leong and Rankin, *Secrets and Knowledge in Medicine and Science.*

37. S.v. "secret," *Dictionnaire de l'Académie françoise*, 2:644.

38. See Eamon, *Science and the Secrets of Nature*; Leong and Rankin, *Secrets and Knowledge in Medicine and Science.*

39. See Chicco, *La seta in Piemonte.*

40. Copies of these regulations are preserved in the ANP, F12/1432A. An anonymous letter to the Bureau dated 1747 mentions a model of Piedmont's spinning machine in the orangerie of the intendant of Languedoc Le Nain (ANP, F12/1453A); on the enticement of skilled workers from Piedmont, see ANP, F12/1432A.

41. Minard, *La fortune du colbertisme*; Meyssonnier, *La balance et l'horloge*; Mukerji, *Impossible Engineering.*

42. ANP, F12/1434 and F12/1435. See also Doyon and Liaigre, *Vaucanson.*

43. The Archives nationales in Paris contain plenty of material on the Jubiés. In 1745 a royal council decree (*arrêt*) granted them eight years to set up an organzine manufacture in Mounteban, in southern France. The French Bureau of Commerce was aware of their role in the manufacture of high-quality organzine and supported their activities, eventually supplying them with the state aid they requested. One of the Jubiés became inspector of manufactures (ANP, F12/1435 and F12/1436). Their overall plan is outlined in "Memoire sur le bon tirage de la soie" (AN F12/ AN F12/1432A). See also Ballot, *Introduction du machinisme dans l'industrie française.*

44. Minard, *La fortune du colbertisme*; Hilaire-Pérez, "Inventing in a World of Guilds"; Kaplan, *Les ventres de Paris.*

45. Doyon and Liaigre, *Jacques Vaucanson.*

46. Vaucanson's compensation for his spinning machine is listed in ANP, F12/823. ANP, F12/821, lists several other generous compensations that Vaucanson received from the Bureau. On the failure of his "great design," see Doyon and Liaigre, *Jacques Vaucanson.*

47. The Jubiés believed that the Bureau should abolish the *petits tirages* (ANP, F12/1432A, "Memoire sur le bon tirage de la soie").

48. On the "affair Soumille," see ANP, F12/2201.

49. ANP, F12/2201. See also Hilaire-Pérez, *L'invention technique au siècle des Lumières.*

50. ANP, F12/1453, "De la méthode dont le piemontois se servent pour tirer le soie qu'il cueillent (1747)."

51. Minard, *La fortune du colbertisme.* On the history of French eighteenth-century political economy, see Meyssonnier, *La balance et l'horloge*, and Vardi, *The Physiocrats.*

52. Lebeau, "Circulations internationales et savoirs d'État au XVIIIe siècle"; Mukerji, *Impossible Engineering.*

53. Demeulenaere-Douyère and Sturdy, *L'enquête du régent.*

54. Fontenelle, "Eloge de M. Rouillé," 183–84.

55. ANP F12/1453, "Observations sur le filage des soies," Gaudin to Trudaine and Jouy, June 30, 1749. Several dossiers in the F12 series in ANP contain reports with annotations on their front page, presumably penned by a member of the Bureau's staff, that indicate that the reports were forwarded to Nollet.

56. ANP, F12/1453.

57. ANP, F12/1453, "Observations sur le filage des soies," Nollet to Trudaine, July 31, 1749.

58. BMS, MS 150, Nollet, "Journal du voyage de Piedmont et d'Italie en 1749," f. 143v.

59. Bacon's views on dissimulation and secrecy are discussed in Snyder, *Dissimulation and the Culture of Secrecy in Early Modern France*, and Shapin, *A Social History of Truth.*

60. On early modern discussions on dissimulation and secrecy, see Snyder, *Dissimulation and the Culture of Secrecy.*

61. ANP, F12/1453, "Observations sur le filage des soies," Nollet to Gaudin, June 21, 1749.

62. Nollet had to refuse this decoration at first; it took several years for him to obtain permission from the French king to accept it (ANP, O1, 402, f. 664v–65, December 29, 1760).

63. On the role of intermediaries and go- betweens, see Schaffer, Roberts, Raj, Delbourgo, and Sibum, *The Brokered World.*

64. ANP, F12/1453, "Observations sur le filage des soies," Nollet to the Bureau, June 21, 1749.

65. On the gentlemanly codes of early modern experimental philosophy see Shapin and Schaffer, *Leviathan and the Air-Pump.*

66. AN, F12/1453, "Observations sur le filage des soies," Nollet to Trudaine, July 31, 1749.

67. In 1747 Matthey obtained a royal privilege for a furnace that allowed owners of silk manufactures to save on wood. He also designed a hygrometer that enabled merchants to measure how much humidity had been absorbed by silk bundles and, most significantly, he invented a machine that would be later called a "serimeter" that allowed merchants to select the best bundles. The serimenter was commissioned by a Genevan banker and caused "a lot of distress and jealousy among the merchants who did not own it" (BMS, MS 150, Nollet, "Journal du voyage de Piemont et d'Italie en 1749," f. 49v). On the privilege that Matthey obtained and his various inventions for silk manufacturers and merchants, see BRT, Storia Patria 1096, Ghiliossi di Lemie, "Arti e Manifatture," f. 111. Matthey's inventions are referred to as "state secrets" by Edward Gibbon; see his *Viaggio in Italia*, 61–62.

68. AST, Reg. 22, ff. 91v–92v.

69. BMS, MS 150, Nollet, "Journal du voyage de Piedmont et d'Italie en 1749," f. 71. Nollet did present Matthey's invention to the Academy; see AASP, PV 71 (1752), f. 389.

70. Matthey belonged to a Protestant minority that held strong ties with the Haldimans, a family of bankers in Geneva who exported Piedmont's organzine. This is documented in a letter in the de Luc papers (BPUG, Haldiman to de Luc, September 6, 1771).

71. Ferrone, *La nuova Atlantide e i lumi.*

72. Gibbon, *Viaggio in Italia*, 59.

73. ANP, F12/1453, "Observations sur le filage des soies," Nollet to Trudaine, June 21, 1749. Most likely this man was one of the Gioanettis, a family of magistrates that owned silk manufactures in Turin. In 1746 the Gioanettis signed a petition to the king to obtain more favorable regulations for entrepreneurs; see Cerutti, *Giustizia sommaria*.

74. ANP, F12/1453B, "Observations sur le filage des soies," Nollet to Trudaine, June 21, 1749.

75. ANP, F12/1453B.

76. On eighteenth-century French political economy, see Meyssonnier, *La balance et l'horloge*, and Vardi, *The Physiocrats and the World of the Enlightenment*.

77. Around 1750, the idea of a spinning academy was elaborated on by Montessuy, *maitre-garde* of the Lyons merchants, who accompanied Vaucanson in Piedmont in 1741 and contributed to his plan; see Doyon and Liagre, *Jacques Vaucanson*.

78. ANP, F12/821. Vaucanson received an annual salary of twelve thousand livres. The Jubiés strongly opposed Vaucanson's spinning machine; see ANP, F12/ 1342A, "Observation sur le tour a filer la soie de Vocanson." This memoir contains a draft of regulations that the Jubiés sent to Trudaine in 1751.

79. ANP, F12/810.

80. ANP, F12/1453B.

81. ANP, F12/810. Nollet paid fifty livres "dans les cassins et ateliers" in Piedmont, but no money is mentioned to obtain the memoirs and drawings from his informants in Turin.

82. On the intendants' per diem, see Minard, *La fortune du Colbertisme*.

83. ANP F12/821. (On Vaucanson's compensation, see ANP, F12/810.).

84. Doyon and Liagre, Jacques *Vaucanson*.

85. ANP, F12/2201.

86. ANP, F12/2201.

87. ANP, F12/1311, Jars to Trudaine, May 1764, quoted in Harris, *Industrial Espionage and Technology Transfer*, 591n12.

88. ANP, F12/1453B.

89. Nollet, "Expériences et observations en différent endroits d'Italie," 466–69.

90. Stewart, "A Meaning for Machines," 263. On diplomatic espionage, see Bély, *Espions et ambassadeurs au temps de Louis XIV*.

91. Guidicini, *Miscellanea storico-patria bolognese*, 267; Poni, "Archéologie de la fabrique," 1480.

92. BMS, MS 150, Nollet, "Journal du voyage de Piemont et d'Italie en 1749," f. 120v.

93. Poni, "Archéologie de la fabrique."

94. Chicco, *La seta in Piemonte*.

95. Hutton, *The History of Derby*.

Chapter 2 · *Electricity, Enlightenment, and Deception*

1. Priestley, *The History and Present State of Electricity*, xv.

2. Voltaire, *The Wordling*.

3. Haller, "An Historical Account of the Wonderful Discoveries Made in Germany, etc., Concerning Electricity." See also Pivati, *Nuovo dizionario scientifico e curioso sacro-profano*, 3, s.v. "elettricità." On how electricity contributed to affirming racial superiority over enslaved Africans, see Delbourgo, *A Most Amazing Scene of Wonders*.

4. Lovett, *The Subtil Medium Prov'd*.

5. Muratori, *Edizione nazionale del carteggio di L. A. Muratori*, 14:335.

6. Rousseau, *The Miscellaneous Works of Mr. J. J. Rousseau*, 86.

7. Krüger, *Naturlehre*, 543.

8. Krüger, *Zuschrift an seine Zuhörer*. On electricity in Germany, see Hochadel "The Sale of Shocks and Sparks."

9. Heering, Hochadel, and Rhees, *Playing with Fire*; Delbourgo, *A Most Amazing Scene of Wonders*; Riskin, *Science in the Age of Sensibility*.

10. Algarotti, *Opere*, 9:85.

11. The work was published anonymously. See also Cavazza, *Settecento inquieto*.

12. Schaffer, "Natural Philosophy and Public Spectacle"; Schaffer, "The Consuming Flame"; Stewart, *The Rise of Public Science*.

13. Haller, "An Historical Account of the Wonderful Discoveries Made in Germany, etc., Concerning Electricity." The *Gentleman's Magazine* was popular among the educated, selling up to ten thousand copies an issue. It covered vast areas of political, social, financial and cultural life (Porter, "Laymen, Doctors and Medical Knowledge").

14. Haller, "An Historical Account of the Wonderful Discoveries Made in Germany, etc., Concerning Electricity."

15. On Gray's experiments, see Schaffer "Experimenters' Techniques, Dyers' Hands, and the Electric Planetarium."

16. On this topic see Gentilcore, *Medical Charlatanism in Early Modern Italy* and Minuzzi, *Sul filo dei segreti*.

17. For an overview of the various electrical theories in the eighteenth century, see Heilbron, *Electricity in the 17th and 18th Centuries*. On Franklin and his Italian followers, see Pace, *Benjamin Franklin and Italy*.

18. On the problem of replication, see Sibum, "Reworking the Mechanical Value of Heat," Schaffer, "Glass Works," and Heering, "The Enlightened Microscope."

19. Priestley, *The History and Present State of Electricity*, xi, 247.

20. On the relationship between philosophical instruments and natural magic, see Hankins and Silverman, *Instruments and the Imagination*.

21. Nollet, *Essai sur l'électricité des corps*, 66–67, 84–85.

22. On dance in early modern France, see Cohen, *Art, Dance, and the Body in French Culture of the Ancien Régime*.

23. On social dances during the ancien régime, see Cohen, *Art, Dance, and the Body in French Culture of the Ancien Régime*.

24. Haller, "An Historical Account of the Wonderful Discoveries Made in Germany, etc., Concerning Electricity," 193–94.

25. Bose, *Recherches sur la cause et sur la véritable théorie de l'électricité*; Bose, *Tentamina electrica tandem aliquando hydraulicae chymiae et vegetalibus utilia pars postior*; Bose, *L'électricité*.

26. Bertucci, "Sparks in the Dark."

27. Bertucci, "Sparks in the Dark."

28. de Castro, "Le fluide électrique chez Sade."

29. Porter, "The Sexual Politics of James Graham."

30. Lynn, *Popular Science and Public Opinion in Eighteenth-Century France*.

31. BNM, It., X, 288, f. 91, Wabst to Poleni, November 30, 1745. The University of Padua covered such expenses; see Lat., VIII, 158, BNM.

32. Maffei, *Epistolario*, 2:1164.

33. Anon., review of *Due dissertazioni della elettricità applicate alla medicina*, by Eusebio Sgaurio, 28.

34. BMN, Ms 138, f. 185v, Bianconi to Séguier, November 11, 1746.

35. Muratori, *Edizione nazionale del carteggio di L. A. Muratori*, 20:344.

36. Altieri Biagi and Basile, *Scienziati del Settecento*.

37. Muratori, *Edizione nazionale del carteggio di L. A. Muratori*, 20:345–46.

38. Muratori, *Edizione nazionale del carteggio di L. A. Muratori*, 20:345–46.

39. Muratori, *Edizione nazionale del carteggio di L. A. Muratori*, 20:345–46.
40. BNM, Lat., VIII, 146.
41. BNM, It., X, 288, f. 94, Poleni to Castelnuovo, March 8, 1747.
42. Maffei, *Epistolario*, 2:1170.
43. Maffei, *Epistolario*, 2:1177; Maffei, *Epistolario*, 2:1178.
44. Maffei, *Epistolario*, 2:1177; Maffei, *Della formazione de' fulmini trattato del sig. marchese Scipione Maffei raccolto da varie sue lettere*, 152.
45. Maffei, *Epistolario*, 2:1177.
46. BCV, cod. DCCCCLXXIV, fasc. IV, Bossaert to Maffei, June 3, 1747.
47. BNM, Lat., VIII, 146.
48. Mariotti, *Lettera scritta ad una dama dal signor dottore Prospero Mariotti sopra la cagione de' fenomeni della macchina elettrica.*
49. Muratori, *Edizione nazionale del carteggio di L. A. Muratori*, 14:334.
50. iStay, *Philosophiae versibus traditae libri vi*, 148. See also Haskell, *Loyola's Bees.*
51. Mazzolari, *Electricorum*, 6n33.
52. Muratori, *Edizione nazionale del carteggio di L. A. Muratori*, 14:335.
53. Nollet, *Lettere intorno all'elettricità*, translator's introduction. The anonymous translator was Mariangela Ardinghelli. See Bertucci, "The In/visible Woman."
54. On artisanal tools repurposed for experiments, see Bertucci, *Artisanal Enlightenment*, and Werrett, *Thrifty Science.*
55. Maffei, *Della formazione de' fulmini trattato del sig. marchese Scipione Maffei raccolto da varie sue lettere*, 152.
56. Maffei, *Della formazione de' fulmini trattato del sig. marchese Scipione Maffei raccolto da varie sue lettere*; BCV, Laboratorio Gazoliano, 1.n.9. On science in eighteenth-century Veneto, see Dal Prete, *Scienza e società nella terraferma veneta*, and Dal Prete, "The Gazola Family's Scientific Cabinet."
57. Mazzuchelli, *Gli scrittori d'Italia*, 1:979. On Ardinghelli, see Bertucci, "The In/visible Woman."
58. Cavazza, "Between Modesty and Spectacle"; Cavazza, "Dottrici e lettrici dell'Università di Bologna nel Settecento"; Findlen, "Science as a Career in Enlightenment Italy."
59. Cavazza, "Laura Bassi e il suo sabinetto di fisica sperimentale"; Cavazza, "Laura Bassi and Giuseppe Veratti." Among those who benefited from the activities in the Bassi-Veratti home were a number of distinguished protagonists of late eighteenth-century scientific life, including Lazzaro Spallanzani, Felice Fontana, and Giambattista Beccaria; see Cavazza, "Laura Bassi 'maestra' di Spallanzani."
60. Bergamini, *Interni d'accademia.*
61. [Zanotti], *Della forza attrattiva delle idee*, 20–21.
62. [Sguario and Wabst], *Dell'elettricismo*, 367, 389.
63. Krüger, *Naturlehre*, 544.
64. Krüger, *Naturlehre*, 46.
65. Krüger, *Der Weltweisheit und Artzneygelahrheit Doctors und Professors auf der Friedrichs Universität Naturlehre*, 543.
66. Kratzenstein, *Abhandlung von dem Electricitat in der Urznenwissenschaft*, 35.
67. Bose, "Abstract of a Letter from Monsieur de Bozes," 420.
68. [Sguario and Wabst], *Dell'elettricismo*, 372.
69. Lovett, *The Subtil Medium Prov'd.* See also Bertucci, "Shocking Subjects."
70. Barry, "Piety and the Patient"; Bertucci, "Shocking Subjects."
71. Wesley, *Primitive Physic*; Wesley, *The Desideratum.* On Wesley's electrical treatments, see Bertucci, "Revealing Sparks."

72. BNM, Lat., VIII, 146.

73. BNM, It., X, 288, f. 89, Poleni to Sale, March 13, 1747; BNM, It., X, 288, f. 91, Sale to Poleni, March 8, 1747; BNM, It., X, 309, f. 21, Bossaert to Poleni, February 21, 1747; BNM, vol. 9352, "Copia della lettera del sig. marchese Luigi Sale di Vicenza al sig. Gian Ludovico Bianconi medico di S.A. il principe e vescovo di Augusta, March 12, 1747"; Cerisara, *Lettera del signor Giambattista Cerisara al signor dottor Giammaria Pigati sopra la convulsione elettrica.*

74. Nollet, "Observations sur quelques nouveaux phénomènes d'électricité," 19.

75. Jallabert, *Expériences sur l'électricité avec quelques conjectures sur la cause de ses effets.*

76. BPUG, Collection Jallabert, MS 82, f. 41, August 15, 1746. On de Sauvages, see Williams, *A Cultural History of Medical Vitalism in Enlightenment Montpellier.* On medical electricity more generally in France, see Zanetti, *L'électricité médicale dans la France des Lumières.*

77. Benguigui, *Théories électriques du XVIIIe siècle,* 64.

78. Mortimer, "A Letter to Martin Folkes, Esq., President of the Royal Society, from Cromwell Mortimer, M. D., Secr. of the Same, Concerning the Natural Heat of Animals," 479.

79. Baker, "A Letter from Mr. Henry Baker, F. R. S., to the President, Concerning Several Medical Experiments of Electricity."

80. [Sguario and Wabst], *Dell'elettricismo,* 383. On Galvani, see Piccolino and Bresadola, *Rane, torpedini e scintille.* On Galvani, see Piccolino and Bresadola, *Shocking Frogs.*

81. [Bianchi], "Della formazione dei fulmini trattato del marchese Scipione Maffei," 654. The authorship of these lines is to be attributed to the physician from Rimini, Giovanni Bianchi: the draft of his review is in the minutes of his letters to Giovanni Lami, the editor of *Novelle letterarie;* see Bianchi to Lami, September 23, 1747, in BGR, SC-MS 970, ff. 82–82v, Bianchi, minute di lettere dal 1745 al 1761.

82. Belgrado, *I fenomeni elettrici con i corollari da lor dedotti.*

83. [Bianchi], "Della formazione dei fulmini trattato del marchese Scipione Maffei," 654.

84. Della Torre, *Elementa Physicae,* 5:330. For a more detailed discussion of the plate, see Bertucci, "Architecture of Knowledge."

85. Nollet, *Recherches sur les causes particulières des phénomènes électriques,* xvii.

86. Nollet, *Recherches sur les causes particulières des phénomènes électriques,* 279.

87. Nollet, *Recherches sur les causes particulières des phénomènes électriques,* 301–2. In urging that only medical practitioners make these decisions, Nollet avoided entering contemporary medical debates on Sanctorian medicine; see Dacome, "Living with the Chair," and Dacome, "Balancing Acts."

88. On the delicate relationship between medical professionals and charlatans, see Gentilcore, *Medical Charlatanism in Early Modern Italy,* Porter, *Quacks,* and Cosmacini, *Il medico saltimbanco.*

89. Nollet, *Leçons de physique expérimentale,* 1:52.

90. Heilbron, *Weighing the Imponderables and Other Quantitative Science around 1800.*

91. Nollet, *Recherches sur les causes particulières des phénomènes électriques,* 247–49.

92. J. W., "Salutary Effects of Electricity," 238.

93. On credibility, testimony, and trust in early modern scientific accounts, see Shapin, *A Social History of Truth.*

94. Review of *Istoria d'un sonnambulo,* by Giovanni Maria Pigatti, 179. The quote from Horace—"in vacuo laetus essor plausorque theatro"—is from a poem about a fool in Argo to whom the reviewers, pretending to believe the marvelous story of the sleep-walker, compared themselves: "Fuit haud ignobilis Argis/Qui se credebat miros audire tragoedos/ In vacuo laetus sessor plausorque theatro" ("There was once a noble man in Argo, who believed that he heard marvelous tragic actors, while he sat happily in an empty theater").

95. Heilbron, *Electricity in the 17th and 18th Centuries.*

96. Werrett, *Fireworks*.

97. Gentilcore, *Medical Charlatanism in Early Modern Italy*; Minuzzi, *Sul filo dei segreti*.

98. Van Swieten, *Vampyrismus*.

99. Winkler, "Novum reique medicae utile electricitatis inventum."

100. Benguigui, *Théories électriques du XVIIIe siècle*, 173.

101. AASP, PV 67 (1748), ff. 486–87.

102. Nollet, *Recherches sur les causes particulières des phénomènes électriques*, 308.

103. Benguigui, *Théories du XVIIIe siècle*, 165.

Chapter 3 · Fabricated Controversy

1. Nollet, "Expériences et observations en différent endroits d'Italie," 444, 445, 446, 452.

2. Nollet, "Extract of a Letter from the Abbé Nollet, F. R. S., etc. to Charles, Duke of Richmond, F. R. S., Accompanying an Examination of Certain Phaenomena in Electricity, Published in Italy," 397.

3. On the love of the marvelous as an outdated passion during the Enlightenment, see Daston and Park, *Wonders and the Order of Nature*.

4. On the persistence of the marvelous, see Evans and Marr, *Curiosity and Wonder from the Renaissance to the Enlightenment*; Ionescu and Smylitopoulos, "Wonder in the Eighteenth Century."

5. Anon., review of *Essai sur l'électricité des corps*; anon., "Abbé Nollet's Examination of Electricity"; anon., "Medical Effects from Electricity"; anon., "Nollet's Examination of Electrical Experiments."

6. Anon., review of *Essai sur l'électricité des corps*.

7. The literature on public opinion is vast and cannot be fully listed here. See at least La Vopa, "The Birth of Public Opinion," and Lynn, *Popular Science and Public Opinion in Eighteenth-Century France*.

8. Darnton, "Policing Writers in Paris circa 1750"; Darnton, "The True History of Fake News"; Soll, "The Long and Brutal History of Fake News."

9. del Negro, "L'Università," 64–65.

10. ASV, Riformatori, f. 10, cc. 157–63, cited in Infelise, *L'editoria veneziana nel '700*, 41.

11. Dooley, "Le accademie"; Dooley, *Science, Politics, and Society*; Berengo, *Giornali veneziani del Settecento*.

12. Berengo, *Giornali veneziani del Settecento*; Berengo, *La società veneta alla fine del Settecento*; Dooley, "Il patriziato veneziano e l'attività accademica"; Dooley, *Science, Politics, and Society*.

13. Pivati, *Della elettricità medica lettera del chiarissimo Signore Gio. Francesco Pivati*

14. Pivati, *Nuovo dizionario scientifico e curioso sacro-profano*, 3:xiv.

15. Pivati, *Nuovo dizionario scientifico e curioso sacro-profano*, 3:xiv. Pivati, *Della elettricità medica lettera del chiarissimo Signore Gio. Francesco Pivati*.

16. BPUG, Correspondance Jallabert, SHAG 242, f. 186, Zanotti to Jallabert, July 1, 1748.

17. Urbinati, "Physica."

18. Pivati, *Della elettricità medica lettera del chiarissimo Signore Gio. Francesco Pivati*, vi.

19. AASB, Minute, MZ, tit. III, n. 235, March 2, 1749.

20. Algarotti, *Opere*, 12:259.

21. BPUG, Correspondance Jallabert, SHAG 242, ff. 185–87, Zanotti to Jallabert, July 1, 1748; AASP, PV, 67 (1748), f. 425.

22. AASP, PV, 67 (1748), ff. 26–27.

23. Zanotti, "De electricitate medica." On Zanotti's control over the journal, see, AASB, Lezioni e carte diverse, tit. IV, fasc. XI, Bonzi, "Notizie e giudizi sulle dissertazioni." Bonzi's

report, which does not survive, was titled "Sopra la medicatura elettrica"; he read it to the Bologna Academy of Sciences on December 7, 1747; see AASB, Registro Atti, published in Angelini, *Anatomie accademiche*, 337. For references to Bonzi's report, see Morgagni, *Carteggio tra Giambattista Morgagni e Francesco M. Zanotti*, 349–51.

24. AASP, PV, 68 (1749), ff. 16–26.

25. AASP, PV, 68 (1749), ff. 16–26.

26. AASP, PV, 68 (1749), ff. 16–26; Bassi, *Lettere inedite alla celebre Laura Bassi scritte da illustri italiani e stranieri*, 119.

27. AASP, PV, 68 (1749), f. 20.

28. BPUG, Correspondance Jallabert, SHAG 242, f. 199v, Garro to Jallabert, March 29, 1749.

29. AASP, PV, 68 (1749), ff. 18–19, Garro to Nollet, November 4, 1748.

30. Muratori, *Edizione nazionale del carteggio di L. A. Muratori*, 20:405.

31. Maffei, *Epistolario*, 2:1239.

32. Bammacaro, *Tentamen de vi electrica ejusque phaenomenis*, 183n.

33. Maffei, *Epistolario*, 2:1239.

34. Bassi, *Lettere inedite alla celebre Laura Bassi scritte da illustri italiani e stranieri*, 115.

35. Findlen, "Science as a Career in Enlightenment Italy"; Cavazza, "Laura Bassi and Giuseppe Veratti."

36. Pivati, *Riflessioni fisiche sulla medicina elettrica*.

37. Pivati, *Riflessioni fisiche sulla medicina elettrica*. On charlatans' strategies, see Gentilcore, *Medical Charlatanism in Early Modern Italy*, Minuzzi, *Sul filo dei segreti*, Havens, "Babelic Confusion," Havens, "Forgery," and Lynch, *Deception and Detection in Eighteenth-Century Britain*.

38. Heilbron, *Electricity in the 17th and 18th Centuries*; Schaffer, "Self Evidence"; Priestley, *History and Present State of Electricity*.

39. Garofalo, *L'enciclopedismo italiano*; Infelise, "Enciclopedie e pubblico a Venezia a metà Settecento."

40. Torrini, "Il gioco e lo svago dei veneziani"; Plebani, "Socialità, conversazioni e casini nella Venezia del secondo Settecento."

41. About ten years later, Querini's campaign to restore the Great Council exploded into a revolt and culminated with Querini's arrest and exile. See Venturi, *Settecento riformatore*, and Brunelli Bonetti, "Un riformatore mancato."

42. BMS, MS 150, Nollet, *Journal du voyage de Piedmont et d'Italie en 1749*, f. 104.

43. BMS, MS 150, Nollet, *Journal du voyage de Piedmont et d'Italie en 1749*, f. 104.

44. Vivian, *Il console Smith mercante e collezionista*; Torcellan, *Un économiste du XVIIIe siècle*; Venturi, *Settecento riformatore*. On Italian translations of Chambers's encyclopedia, see Farinella, "Le traduzioni italiane della *Cyclopedia* di Ephraim Chambers."

45. BGR, Fondo Gambetti, Pasquali to Bianchi, November 26, 1747.

46. Vivian, *Il console Smith mercante e collezionista*, 100–12.

47. Pasquali published the Italian translations of Nollet's *Essai sur l'électricité des corps*, *Leçon de physique experimentale*, and *Recherches sur les causes particulieres de phénoménes électriques*. [particulières?]

48. Bianchini, *Saggio d'esperienze intorno alla medicina elettrica fatte in Venezia da alcuni amatori di fisica al Signor Abate Nollet*; Watson, "An Account of Dr. Bianchini's *Recueil d'experiences faites à Venise sur la medicine électrique*."

49. BMS, MS 150, Nollet, "Journal du voyage de Piedmont et d'Italie en 1749," f. 101v.

50. Nollet, "Expériences et observations en différent endroits d'Italie " 444, 452.

51. Benguigui, *Théories électriques du XVIIIe siècle*, 174.

52. Benguigui, *Théories électriques du XVIIIe siècle*, 176.

53. Benguigui, *Théories électriques du XVIIIe siècle*, 177.
54. Benguigui, *Théories électriques du XVIIIe siècle*, 180.
55. Bassi, *Lettere inedite alla celebre Laura Bassi scritte da illustri italiani e stranieri*, 194.
56. BCAB, Collezione autografi, LVI, f. 15002, Pivati to Veratti, August 2, 1749. .
57. BCAB, Collezione autografi, LVI, f. 15001, Pivati to Veratti, July 1749.
58. Benguigui, *Théories électriques du XVIIIe siècle*, 165.
59. BCAB, B 160, lettere di vari a Zanotti, f. 38, Nollet to Zanotti, October 29, 1749.
60. BCAB, B 160, lettere di vari a Zanotti, f. 35, Nollet to Zanotti, September 6, 1749.
61. BMS, MS 150, Nollet, "Journal du voyage de Piedmont et d'Italie en 1749," f. 91.
62. Benguigui, *Théories électriques du XVIIIe siècle*, 177–79.
63. BMS, MS 150, Nollet, "Journal du voyage de Piedmont et d'Italie en 1749," f. 116v.
64. Vedova, *Biografia degli scrittori padovani*; Messbarger, *The Century of Women*.
65. Badinter and Muzurelle, *Madame u Châtelet*; Zinsser, "The Many Representations of the Marquise Du Châtelet."
66. BMS, MS 150, Nollet, "Journal du voyage de Piedmont et d'Italie en 1749," f. 115.
67. BMS, MS 150, Nollet, "Journal du voyage de Piedmont et d'Italie en 1749, " f. 115.
68. Benguigui, *Théories électriques du XVIIIe siècle*, 177–78.
69. Kraus, *Briefe Benedicts XIV an Den Canonicus Pier Francesco Peggi in Bologna*.
70. BCAB, B 160, lettere di vari a Zanotti, f. 45, Nollet to Zanotti, May 11, 1750.
71. RSL, journal book, 1750, ff. 278–86.
72. Bassi, *Lettere inedite alla celebre Laura Bassi scritte da illustri italiani e stranieri*, 221. For Veratti's draft reply, see BCAB, MSS Laura Bassi, cart. 1, fasc. 2.
73. AASB, minute, tit. III, Zanotti to Nollet, December 19, 1750. In a letter to Veratti, dated July 10, 1751, Pivati shared his disappointment at Veratti's lack of interest in the medicated tubes; see BCAB, Collezione autografi, LVI, f. 15004.
74. Bertucci, "Enlightening Towers"; Bresadola and Piccolino, *Shocking Frogs*; Pancaldi, *Volta*; Delbourgo, *A Most Amazing Scene of Wonders*.
75. Zanetti, "Curing with Machines"; Bertucci, "Shocking Subjects."
76. Nollet, "Expériences et observations en différent endroits d'Italie," 448; BMS, MS 150, Nollet, *Journal du voyage de Piedmont et d'Italie en 1749*, f. 19.
77. Schaffer, "Self Evidence," 339.
78. Nollet, "Expériences et observations en différens endroits d'Italie," 448; Schaffer, "Self Evidence."
79. Nollet, "Expériences et observations en différens endroits d'Italie," 451.
80. BMS, MS 150, Nollet, "Journal du voyage de Piedmont et d'Italie en 1749," f. 19.
81. Louis, *Observations sur l'électricité*.
82. Nollet, "Expériences et observations en différens endroits d'Italie," 452.
83. Nollet, "Expériences et observations en différens endroits d'Italie," 453.
84. Nollet, "Expériences et observations en différens endroits d'Italie," 454.
85. Nollet, "Expériences et observations en différens endroits d'Italie," 457.
86. Spector, "The 'Lights' before the Enlightenment"; La Vopa, "The Birth of Public Opinion."
87. Algarotti, *Opere*, 9:292–96.
88. On the long-term history of medical electricity, see Bertucci and Pancaldi, *Electric Bodies*.
89. Bassi, *Lettere inedite alla celebre Laura Bassi scritte da illustri italiani e stranieri*, 121–22; Melli, *Studi e inediti per il primo centenario dell'Istituto Magistrale "Laura Bassi,"* 140.
90. Bar-Ilan and Halevi, "Retracted Articles."
91. Darnton, "Policing Writers in Paris circa 1750."
92. Natale, *Gli specchi della paura*.
93. See Carnelos, "Libri da grida da banco e da bottega," for Moneti's quotation and references to the Riformatori dello studio di Padova.

94. Bycroft, "Wonders in the Academy"; Lynch, *Deception and Detection in Eighteenth-Century Britain*; Donato, *The Life and Legend of Catterina Vizzani*.

95. Several similar cases are mentioned in Darnton, *Mesmerim and the End of the Enlightenment in France*. On Tofts, see Cody, *Birthing the Nation*. On de la Plantades's hoax, see Darnton, *Mesmerism and the End of the Enlightenment in France*, and Ruestow, "Leeuwenhoek and the Campaign against Spontaneous Generation." On eighteenth-century theories of generation, see Dal Prete, "Cultures and Politics of Preformationism in Eighteenth-Century Italy."

96. Füssel, "'The Charlatanry of the Learned.'"

97. Daston, "The Ethos of Enlightenment."

98. Priestley, *History and Present State of Electricity*, 10. On early modern debates on history and its functions, see Grafton, *What Was History?*

Chapter 4 · *Natural Marvels, Instruments, and Stereotypes*

1. Vaccari, "The Organized Traveller"; Rubiés, "Instructions for Travellers."

2. The literature on naturalistic journeys is mostly in Italian. See Ciancio, "Alberto Fortis e la pratica del viaggio naturalistico," Ciancio, *Le colonne del tempo*, Di Mitri and Fantini, *La febbre del viaggio*, Cavazza, "'Philocentria' e Pietramala," Cioffi and Martelli, *La Campania e il Grand Tour*, Scataglini, *Tutt'intorno Roma*, Valensise, "Impressioni di viaggio nella Calabria Ulteriore dal diario di Dominique Vivant Denon," Bossi and Greppi, *Viaggi e scienza*, Vaccari, "The Organized Traveller," Carozzi, *Horace-Bénédict de Saussure*, Elsner and Rubiés, *Voyages and Visions*, and Cocco, *Watching Vesuvius*, chap. 7.

3. Goethe, *Goethe's Letters from Switzerland and Travels in Italy*.

4. Prosperi, "Otras Indias." See also Cooley, "Southern Italy and the New World in the Age of Encounters."

5. Calaresu, "From the Street to Stereotype"; Mozzillo, *La frontiera del Grand Tour*; Mozzillo, *Passaggio a Mezzogiorno*.

6. On the influence of Montesquieu's climate theories on the representation of Neapolitans in travel narratives, see Calaresu, "Looking for Virgil's Tomb," 344.

7. Cocco, *Watching Vesuvius*.

8. Nollet, "Expériences et observations en différent endroits d'Italie," 444. On philosophers' indifference to marvels, see Daston and Park, *Wonders and the Order of Nature*, chap. 7.

9. Terrall, *Catching Nature in the Act*.

10. Calaresu, "From the Street to Stereotype."

11. Bossi and Greppi, *Viaggi e scienza*; Carozzi, *Horace-Bénédict de Saussure*; Elsner and Rubiés, *Voyages and Visions*.

12. Prosperi, "Otras indias." See also Cooley, "Southern Italy and the New World in the Age of Encounters," and La Condamine, "Extrait d'un journal d'un voyage en Italie," 388, 391. On La Condamine's journey to South America, see Safier, *Measuring the New World*.

13. Brydone, *A Tour through Sicily and Malta, in a Series of Letters to William Beckford*, 1:165.

14. Rousseau, *The Miscellaneous Works of Mr. J. J. Rousseau*, 4:85.

15. Markey, *Renaissance Invention*.

16. Schaffer, "Traveling Machines and Colonial Times"; Davison, *The Unforgiving Minute*.

17. La Condamine, "Extrait d'un journal d'un voyage en Italie."

18. Schaffer, "Traveling Machines and Colonial Times"; Davison, *The Unforgiving Minute*. Perkins, *The Reform of Time*.

19. La Condamine, "Extrait d'un journal d'un voyage en Italie," 389–391.

20. Brilli, *Quando viaggiare era un'arte*.

21. Brilli, *Quando viaggiare era un'arte*.

22. Misson, *Nouveau Voyage d'Italie de Monsieur Misson*, 63–66. The mysterious vapor that killed people and small animals was carbon dioxide, which being denser than air sinks to the ground of caves in volcanic areas.

23. Nollet, "Suite des expériences et observations en différens endroits d'Italie," 81. On the connections between archeology and natural history, see Ciancio, *Le colonne del tempo*. On the jokes of nature, the classic work is Findlen, "Jokes of Nature and Jokes of Knowledge."

24. Della Torre, *Storia e fenomeni del Vesuvio*, 6–7; Schettino, "L'insegnamento della fisica sperimentale a Napoli nella seconda metà del Settecento."

25. BMS, MS 150, Nollet, "Journal du voyage de Piedmont et d'Italie en 1749."

26. Petrella, *L'officina del geografo*; Nollet "Suite des expériences et observations en différens endroits d'Italie, 98. On Annius, see Grafton, "Invention of Traditions and Traditions of Invention in Renaissance Europe."

27. Nollet, "Suite des expériences et observations en différens endroits d'Italie," 98.

28. BMS, MS 150, Nollet, "Journal du voyage de Piedmont et d'Italie en 1749," ff. 160–61.

29. Nollet, "Suite des expériences et observations en différens endroits d'Italie," 101.

30. Nollet, "Suite des expériences et observations en différens endroits d'Italie," 92ff.

31. Nollet, "Suite des expériences et observations en différens endroits d'Italie," 92ff.

32. On self-experimentation, see Schaffer "Self Evidence." My interpretation differs with respect to Nollet's tour but not on the general point of the power of the body of the experimenter as provider of evidence.

33. Nollet, 'Extract of the Observations Made in Italy," 55.

34. La Condamine, "Extrait d'un journal d'un voyage en Italie," 370–71.

35. Shapin and Schaffer, *Leviathan and the Air-Pump*.

36. BMS, MS 150, Nollet, "Journal du voyage de Piedmont et d'Italie en 1749," f. 156v.

37. Richard de Saint-Non, quoted in Calaresu, "Looking for Virgil's Tomb," 146.

38. Calaresu, "From the Street to Stereotype."

39. BMS, MS 150, Nollet, "Journal du voyage de Piedmont et d'Italie en 1749," f. 176.

40. Fino, *The Miracle of San Gennaro*. On the prefiche, see Cardone, *Riti e rituali a Napoli, in Campania e in Sud Italia*.

41. Kreiser, *Miracles, Convulsions, and Ecclesiastical Politics in Early Eighteenth-Century Paris*.

42. Montesquieu and Montesquieu, *Voyages de Montesquieu*, 2:22–24.

43. Brosses and Babou, *Lettres familières écrites d'Italie à quelques amis en 1739 et 1740*, 374.

44. BMS, MS 150, Nollet, "Journal du voyage de Piedmont et d'Italie en 1749," f. 172.

45. BMS, MS 150, Nollet, "Journal du voyage de Piedmont et d'Italie en 1749," ff. 173r–74.

46. Haas, "Miracles on Trial."

47. BMS, MS 150, Nollet, "Journal du voyage de Piedmont et d'Italie en 1749," ff. 173r–174.

48. BMS, MS 150, Nollet, "Journal du voyage de Piedmont et d'Italie en 1749," f. 17. On Di Sangro and alchemy, see Donato, "Between Myth and Archive, Alchemy and Science in Eighteenth-Century Naples."

49. BMS, MS 150, Nollet, "Journal du voyage de Piedmont et d'Italie en 1749," ff. 174v–75.

50. BMS, MS 150, Nollet, "Journal du voyage de Piedmont et d'Italie en 1749," f. 175.

51. Cocco, *Watching Vesuvius*.

52. BMS, MS 150, Nollet, "Journal du voyage de Piedmont et d'Italie en 1749," f. 175.

53. BMS, MS 150, Nollet, "Journal du voyage de Piedmont et d'Italie en 1749," f. 181.

54. Zanotti, *Della forza de' corpi che chiamano viva libri tre*.

55. BMS, MS 150, Nollet, "Journal du voyage de Piedmont et d'Italie en 1749," f. 175.

56. Bertucci, "The In/Visible Woman"; Vitrioli, *Elogio di Angela Ardinghelli, napoletana*; Findlen, "Translating the New Science."

57. BMS, MS 150, Nollet, "Journal du voyage de Piedmont et d'Italie en 1749," f. 176–176v.

58. Tanucci, *Epistolario*, 499. On ancien régime France as the new Rome, see Mukerji, *Impossible Engineering*. On the transition from Latin to French as the language of the Republic of Letters, also Goodman, *The Republic of Letters*.

59. La Vopa, "Herder's Publikum."

60. BCAB, Collezione autografi, LVI, f. 15002, Pivati to Veratti, August 2, 1749.

61. Calaresu, "Looking for Virgil's Tomb," 151.

62. Della Torre, *Storia e fenomeni del Vesuvio*.

63. Signorelli, *Vicende della coltura nelle due Sicilie*, 494–95.

64. Signorelli, *Vicende della coltura nelle due Sicilie*, 494–95.

65. On reciprocity in the Republic of Letters, see Goodman, *The Republic of Letters*.

Conclusion

1. BMS, Ms 150, "Journal d'un voyage en Italie de l'abbé Nollet en 1749," ff. 194–95, 210–12.

2. BMS, Ms 150, "Journal d'un voyage en Italie de l'abbé Nollet en 1749," ff. 194v, 196, 215.

3. BCAB, B 160, lettere di vari a Zanotti, Nollet to Zanotti, May 11, 1750, B 160, f. 45; AASP, PV, 69 (1750), f. 187. On Ardinghelli as an informal foreign correspondent for the academy, see Bertucci, "The In/visible Woman."

4. Quoted in Daston and Park, *Wonders and the Order of Nature*, chap. 9.

5. Todd, *Imagining Monsters*; Cooper, *Inventing the Indigenous*.

6. On trustworthiness in early modern experimental philosophy, see Shapin, *A Social History of Truth*.

Angelini, Annarita, ed. *Anatomie accademiche*. Vol. 3, *L'Istituto delle scienze e l'accademia*. Bologna: Il Mulino, 1987.

Anon. "Gabinetto di fisica." *Calendario generale pei regii stati* 12 (1835): 532–34.

Anon. "Abbé Nollet's Examination of Electricity." *Gentleman's Magazine* 21 (1751): 261, 349–51.

Anon. "Medical Effects from Electricity." *London Magazine and Monthly Chronologer* 20 (1751): 95–98.

Anon. "Nollet's Examination of Electrical Experiments." *Scots Magazine* 13 (1751): 433–37.

Anon. Review of *Essai sur l'electricité des corps*, by Jean-Antoine Nollet. *Mémoires des Trévoux* 2 (1752): 864–75.

Anon. Review of *Due dissertazioni della elettricità applicata alla medicina*, by Eusebio Sgaurio. *Novelle della repubblica letteraria* (January 1747): 26–28.

Anon. Review of *Essai sur l'électricité des corps*, by Jean-Antoine Nollet. *Universal Librarian* (April–June 1751): 161–65.

Anon. *Traité des ambassades et des ambassadeurs*. Rotterdam: Jean Hofhout, 1726.

Alder, Ken. "History's Greatest Forger: Science, Fiction, and Fraud along the Seine." *Critical Inquiry* 30, no. 4 (2004): 702–16.

Algarotti, Franceso. *Opere*. 17 vols. Venezia: Palese, 1794.

Altieri Biagi, Maria Luisa, and Bruno Basile, eds. *Scienziati del Settecento*. Milan: Ricciardi, 1983.

Badinter, Elisabeth, and Danielle Muzurelle, eds. *Madame Du Châtelet: La femme des Lumières*. Paris: Bibliothèque nationale de France, 2006.

Baker, Henry. "A Letter from Mr. Henry Baker, F. R. S., to the President, Concerning Several Medical Experiments of Electricity." *Philosophical Transactions of the Royal Society* 45 (1748): 270–75.

Ballot, Charles. *L'Introduction du machinisme dans l'industrie française*. Paris: Rieder, 1923.

Bammacaro, Niccolò. *Tentamen de vi electrica ejusque phaenomenis*. Naples: Alexium Pellecchia, 1748.

Baretti, Giuseppe Marc'Antonio. *An Account of the Manners and Customs of Italy, with Observations on the Mistakes of Some Travelers*. 2 vols. London: Davies, 1768–69.

Bar-Ilan, Judit, and Gali Halevi. "Retracted Articles: The Scientific Version of Fake News." In *The Psychology of Fake News: Accepting, Sharing, and Correcting False Information*, edited by Rainer Greifeneder, Mariela E. Jaffé, Eryn J. Newman, and Norbert Schwarz, 47–70. New York: Routledge, 2020.

Barry, Jonathan. "Piety and the Patient: Medicine and Religion in Eighteenth-Century Bristol." In *Patients and Practitioners: Lay Perceptions of Medicine in Pre-Industrial Society*, edited by Roy Porter, 145–76. Cambridge: Cambridge University Press 1985.

Bassi, Laura. *Lettere inedite alla celebre Laura Bassi scritte da illustri italiani e stranieri.* Bologna: Cenerelli, 1885.

Beaurepaire, Pierre-Yves, and Pierrick Pourchasse, eds. *Les circulation internationales en Europe: Années 1680–années 1780.* Rennes: Presses Universitaires de Rennes, 2010.

Belgrado, Jacopo. *I fenomeni elettrici con i corollari da lor dedotti.* Parma: Rosati, 1749.

Bély, Lucien. *Espions et ambassadeurs au temps de Louis XIV.* Paris: Fayard, 1990.

Benguigui, Isaac. *Théories électriques du XVIIIe siècle: Correspondence entre l'abbé Nollet (1700–1770) et le physicien genevois Jean Jallabert (1712–1768).* Geneva: Georg, 1984.

Bensaude-Vincent, Bernadette, and Christine Blondel. *Science and Spectacle in the European Enlightenment.* Aldershot, UK: Ashgate, 2008.

Berengo, Marino, ed. *Giornali veneziani del Settecento.* Milan: Feltrinelli, 1962.

———. *La società veneta alla fine del Settecento: Ricerche storiche.* Florence: Sansoni, 1956.

Beretta, Marco, Antonio Clericuzio, and Lawrence Principe, eds. *The Accademia del Cimento and Its European Context.* Sagamore Beach, MA: Watson, 2009.

Beretta, Marco, and Maria Conforti, eds. *Fakes!? Hoaxes, Counterfeits, and Deception in Early Modern Science.* Sagamore Beach, MA: Watson, 2014.

Bergamini, Maria Grazia. *Interni d'accademia: Il sodalizio bolognese dei Vari, 1747–1763.* Modena: Mucchi, 1996.

Bertrand, Gilles. *Le Grand Tour revisité: Pour une archéologie du tourisme.* Rome: Ecole française de Rome, 2008.

Bertucci, Paola. "The Architecture of Knowledge: Science, Collecting, and Display in Eighteenth-Century Naples." In *New Approaches to Naples, c. 1500–1800,* edited by Melissa Calaresu and Helen Hills, 149–74. Aldershot, UK: Ashgate, 2013.

———. *Artisanal Enlightenment: Science and the Mechanical Arts in Old Regime France.* New Haven, CT: Yale University Press, 2017.

———. "Enlightened Secrets: Silk, Industrial Espionage, and Intelligent Travel in Eighteenth-Century France." *Technology and Culture* 54, no. 4 (2013): 820–52.

———. "Enlightening Towers: Public Opinion, Local Authorities, and the Reformation of Meteorology in Eighteenth Century Italy." *Transactions of the American Philosophical Society* 99 (2009): 25–44.

———. "The In/Visible Woman: Mariangela Ardinghelli and the Circulation of Knowledge between Paris and Naples in the Eighteenth Century." *Isis* 104, no. 2 (2013): 226–49.

———. "Revealing Sparks: John Wesley and the Religious Utility of Electrical Healing." *British Journal for the History of Science* 39, no. 3 (2006): 341–62.

———. "Shocking Subjects: Human Experiments and the Material Culture of Medical Electricity in Eighteenth-Century England." In *The Uses of Humans in Experiment: Perspectives from the 17th to the 20th Century,* edited by Erika Dyck and Larry Stewart, 111–38. Leiden: Brill, 2016.

———. "Sparks in the Dark: The Attraction of Electricity in the Eighteenth Century." *Endeavour* 31, no. 3 (2007): 88–93.

———. "Spinners' Hands, Imperial Minds: Migrant Labor, Embodied Expertise, and the Failed Transfer of Silk Technology across the Atlantic." *Technology and Culture* 60, no. 4 (2021): 1003–31.

———. *Viaggio nel paese delle meraviglie: Scienza e curiosità nell'Italia del Settecento.* Turin: Bollati Boringhieri, 2007.

Bertucci, Paola, and Olivier Courcelle. "Artisanal Knowledge, Expertise, and Patronage in Early Eighteenth-Century Paris: The Société des Arts (1728–36)." *Eighteenth-Century Studies* 48, no. 2 (2015): 159–79.

Bertucci, Paola, and Giuliano Pancaldi, eds. *Electric Bodies: Episodes in the History of Medical Electricity.* Bologna: Università di Bologna, 2001.

[Bianchi, Giovanni]. "Della formazione dei fulmini trattato del marchese Scipione Maffei." *Novelle letterarie pubblicate in Firenze* 8 (1747): 648–56.

Bianchini, Francesco. "Extrait des observations faites au mois de Décembre 1705 par M. Bianchini, sur des feux qui se voient sur une des montagnes de l'Apennin." *Mémoires de l'Académie royale des sciences de Paris* (1706): 336–39.

Bianchini, Gianfortunato. *Saggio d'esperienze intorno alla medicina elettrica fatte in Venezia da alcuni amatori di fisica al Signor Abate Nollet*. Venice: Giambattista Pasquali, 1749.

Blair Ann, Paul Duguid, Anja-Silvia Goeing, and Anthony Grafton, eds., *Information: A Historical Companion*. Princeton, NJ: Princeton University Press, 2021.

Bose, Georg Matthias. "Abstract of a Letter from Monsieur de Bozes." *Philosophical Transactions of the Royal Society* 43 (1744–45): 419–21.

———. *L'électricité: Son origine et ses progrès, poème en deux livres*. Leipzig: Laukisch, 1754.

———. *Recherches sur la cause et sur la véritable théorie de l'électricité*. Wittenberg: Slomac, 1745.

———. *Tentamina electrica tandem aliquando hydraulicae chymiae et vegetalibus utilia pars posterior*. Wittenberg: Ahlfeldium, 1747.

Bossi, Maurizio, and Claudio Greppi, eds. *Viaggi e scienza: Le istruzioni scientifiche per i viaggiatori nei secoli XVII–XIX*. Florence: Olschki, 2005.

Bourget, Marie-Noëlle, Christian Licoppe, and H. Otto Sibum, eds. *Instruments, Travel, and Science: Itineraries of Precision from the Seventeenth to the Twentieth Century*. New York: Routledge, 2002.

Bresadola, Marco, and Marco Piccolino. *Shocking Frogs: Galvani, Volta, and the Electric Origins of Neuroscience*. Oxford: Oxford University Press, 2013.

Bret, Patrice, Irina Gouzevitch, and Liliane Hilaire-Pérez, eds. *Les techniques et la technologie entre France et la Grande-Bretagne: XVIIe-XIXe siècles*. Paris: CNRS, 2010.

Brilli, Attilio. *Il grande racconto del viaggio in Italia: Itinerari di ieri per viaggiatori di oggi*. Bologna: Il Mulino, 2014.

———. *Quando viaggiare era un'arte: Il romanzo del Grand tour*. Bologna: Il Mulino, 1995.

Brizay, François. *Touristes du Grand Siècle: Le voyage d'Italie au XVIIe siècle*. Paris: Belin, 2006.

Brosses, Charles de, and Hippolyte Babou. *Lettres familières écrites d'Italie à quelques amis en 1739 et 1740*. Paris: Poulet-Malassis et de Broise, 1858.

Brunelli Bonetti, Bruno T. "Un riformatore mancato: Angelo Querini." *Archivio Veneto* 81, nos. 83–84 (1951): 185–200.

Brydone, Patrick. *A Tour through Sicily and Malta, in a Series of Letters to William Beckford*. 2 vols. Dublin: James Potts, 1773.

Butterworth, Emily. *Poisoned Words: Slander and Satire in Early Modern France*. Leeds, UK: Legenda, 2007.

Bycroft, Michael. "Wonders in the Academy: The Value of Strange Facts in the Experimental Research of Charles Dufay." *Historical Studies in the Natural Sciences* 43, no. 3 (2013): 334–70.

Calaresu, Melissa. "From the Street to Stereotype: Urban Space, Travel and the Picturesque in Late Eighteenth-Century Naples." *Italian Studies* 62, no. 2 (2007): 189–203.

———. "Looking for Virgil's Tomb: The End of the Grand Tour and the Cosmopolitan Ideal in Europe." In *Voyages and Visions: Towards a Cultural History of Travel*, edited by Jaś Elsner and Joan Pau Rubiés, 138–61. London: Reaktion, 1999.

Calhoun, Craig, ed. *Habermas and thee Public Sphere*. Cambridge, MA: MIT Press, 1992.

Caracciolo, Carlos Héctor. "Notizie false e pratiche editoriali negli avvisi a stampa di antico regime." *L'Archiginnasio: Bulletino della Biblioteca communale di Bologna* 96 (2001): 95–150.

Cardone, Giovanni. *Riti et rituali a Napoli, in Campania, e in Sud Italia*. Rome: Prosspettiva, 2019.

Carnelos, Laura. "Libri da grida, da banco e da bottega: Editoria di consumo a Venezia tra norma e contraffazione (XVII–XVIII)." PhD diss., Università Ca' Foscari Venezia, 2009.

Carozzi, Albert V. *Horace-Bénédict de Saussure (1740–1799): Un pionnier des sciences de la terre*. Geneva: Slatkine, 2005.

Casillo, Robert. *The Empire of Stereotypes: Germaine de Staël and the Idea of Italy*. New York: Palgrave Macmillan, 2006.

Cavazza, Marta. "Between Modesty and Spectacle: Women and Science in Eighteenth-Century Italy." In *Italy's Eighteenth Century: Gender and Culture in the Age of the Grand Tour*, edited by Paula Findlen, Wendy Wassyng Roworth, and Catherine M. Sama, 275–302. Stanford, CA: Stanford University Press, 2007.

———. "Dottrici e lettrici dell'Università di Bologna nel Settecento." *Annali di storia delle università italiane* 1 (1997): 109–26.

———. "Laura Bassi and Giuseppe Veratti: An Electric Couple during the Enlightenment." *Contributions to Science* 5, no. 1 (2009): 115–28.

———. "Laura Bassi e il suo gabinetto di fisica sperimentale: Realtà e mito." *Nuncius* 10 (1995): 715–53.

———. "Laura Bassi 'maestra' di Spallanzani." In *Il cerchio della vita: Materiali di ricerca del Centro studi Lazzaro Spallanzani di Scandiano sulla storia della scienza del Settecento*, edited by Water Bernardi and Paola Manzini, 185–202. Florence: Olschki, 1999.

———. "'Philocentria' e Pietramala: Lessing tra curiosità filosofica e passione bibliofila." In *Gotthold Ephraim Lessing e i suoi contemporanei in Italia*, edited by Lea Ritter Santini, 113–28. Naples: Vivarium, 1997.

———. *Settecento inquieto: Alle origini dell'Istituto delle scienze di Bologna*. Bologna: Il Murino, 1990.

Ceccarelli, Leah. "Manufactured Scientific Controversy: Science, Rhetoric, and Public Debate." *Rhetoric and Public Affairs* 14, no. 2 (2011): 195–228.

Cerisara, Giambattista. *Lettera del signor Giambattista Cerisara al signor dottor Giammaria Pigati sopra la convulsione elettrica*. Vicenza: Berno, 1748.

Cerutti, Simona. *Giustizia sommaria: Pratiche e ideali di giustizia in una società di Ancien Régime (Torino XVIII secolo)*. Milan: Feltrinelli, 2003.

Chapron, Emmanuelle. "Voyageurs et bibliothèques dans l'Italie du XVIIIe siècle des *Mirabilia* au débat sur l'utilité publique." *Bibliothèque de l'Ecole des chartes* 162, no. 2 (2004): 455–82.

Chicco, Giuseppe. *La seta in Piemonte, 1650–1800: Un sistema industriale d'Ancien Régime*. Milan: Angeli, 1995.

Ciancio, Luca. "Alberto Fortis e la pratica del viaggio naturalistico: Stile di ricerca e modalità di prova." *Nuncius* 10, no. 2 (1995): 617–44.

———. *Le colonne del tempo: Il "Tempio di Serapide" a Pozzuoli nella storia eella geologia, dell'archeologia e dell'arte (1750–1900)*. Florence: Edifir, 2009.

Cioffi, Rosanna, and Sebastiano Martelli, eds. *La Campania e il Grand Tour: Immagini, luoghi e racconti di viaggio tra Settecento e Ottocento*. Rome: L'Erma di Bretschneider, 2015.

Cocco, Sean. *Watching Vesuvius: A History of Science and Culture in Early Modern Italy*. Chicago: University of Chicago Press, 2012.

Cody, Lisa Forman. *Birthing the Nation: Sex, Science, and the Conception of Eighteenth-Century Britons*. Oxford: Oxford University Press, 2005.

Cohen, Sarah R. *Art, Dance, and the Body in French Culture of the Ancien Régime*. Cambridge: Cambridge University Press, 2000.

Condamine, Charles Marie de la. "Extrait d'un journal d'un voyage en Italie." *Mémoires de l'Académie royale des sciences de Paris* (1757): 336–410.

Cooley, Mackenzie. "Southern Italy and the New World in the Age of Encounters." In *The New World in Early Modern Italy, 1492–1750*, edited by Elizabeth Hordowich and Lia Markey, 169–89. Cambridge: Cambridge University Press, 2017.

Cooper, Alix. *Inventing the Indigenous: Local Knowledge and Natural History in Early Modern Europe*. Cambridge: Cambridge University Press, 2007.

Cosmacini, Giorgio. *Il medico saltimbanco: Vita e avventure di Buonafede Vitali, giramondo instancabile, chimico di talento, istrione di buona creanza*. Rome: Laterza, 2008.

Dacome, Lucia. "Balancing Acts: Picturing Perspiration in the Long Eighteenth Century." *Studies in History and Philosophy of Biological and Biomedical Sciences* 43, no. 2 (2012): 379–91.

———. "Living with the Chair: Private Excreta, Collective Health, and Medical Authority in the Eighteenth Century." *History of Science* 39, no. 4 (2001): 467–500.

Dal Prete, Ivano. "Cultures and Politics of Preformationism in Eighteenth-Century Italy." In *The Secrets of Generation: Reproduction in the Long Eighteenth Century*, edited by Ray Stephenson and Darren Wagner, 59–78. Toronto: University of Toronto Press, 2015.

———. "The Gazola Family's Scientific Cabinet: Politics, Society, and Scientific Collecting in the Twilight of the Republic of Venice." In *Cabinets of Experimental Philosophy in Eighteenth-Century Europe*, edited by Jim Bennett and Sofia Talas, 155–72. Leiden: Brill, 2013.

———. *Scienza e società nella terraferma veneta: Il caso veronese, 1680–1796*. Milan: Angeli, 2008.

Darnton, Robert. "An Early Information Society: News and the Media in Eighteenth-Century Paris." *American Historical Review* 105, no. 1 (2000): 1–35.

———. *Forbidden Best-Sellers of Pre-Revolutionary France*. New York: Norton, 1995.

———. *Mesmerism and the End of the Enlightenment in France*. Cambridge, MA: Harvard University Press, 1968.

———. "Policing Writers in Paris circa 1750." *Representations* 5 (1984): 1–31.

———. "The True History of Fake News." *New York Review of Books*. February 13, 2017. https://www.nybooks.com/daily/2017/02/13/the-true-history-of-fake-news.

Daston, Lorraine. "The Ethos of Enlightenment." In *The Sciences in Enlightened Europe*, edited by William Clark, Jan Golinski, and Simone Schaffer, 495–504. Chicago: University of Chicago Press, 1999.

Daston, Lorraine, and Katharine Park. *Wonders and the Order of Nature, 1150–1750*. New York: Zone, 1998.

Daumas, Maurice. *Scientific Instruments of the Seventeenth and Eighteenth Centuries*. New York: Praeger Publishers, 1972.

Davies, Surekha. *Renaissance Ethnography and the Invention of the Human: New Worlds, Maps and Monsters*. Cambridge: Cambridge University Press, 2016.

Davini, Roberto. "Bengali Raw Silk, the East India Company and the European Global Market, 1770–1833." *Journal of Global History* 4, no. 1 (2009): 57–79.

Davison, Graeme. *The Unforgiving Minute: How Australia Learned to Tell the Time*. New York: Oxford University Press, 1993.

De Castro, Clara Carnicero. "Le fluide électrique chez Sade." *Dix-huitième siècle* 46, no. 1 (2014): 561–77.

de Clercq, Peter R. *At the Sign of the Oriental Lamp: The Musschenbroek Workshop in Leiden, 1660–1750*. Rotterdam: Erasmus, 1997.

de Fouchy, Jean-Paul Grandjean. "Eloge de M. L'Abbé Nollet." *Histoire de l'Académie royale des sciences* (1770): 121–36.

de Mairobert, Mathieu François Pidanzat. *L'espion anglois; ou, Correspondance secrete entre milord All'eye et milord Alle'ar*. London: John Adamson, 1779.

Demeulenaere-Douyère, Christiane, and David J. Sturdy, eds. *L'enquête du régent, 1716–1718: Sciences, techniques et politique dans la France pré-industrielle.* Turnhout: Brepols, 2008.

de Refuge, Eustache. *Traité de la cour; ou, Instruction des courtisans.* Leide: Elseviers, 1649.

de Rouquefort, Jean-Baptiste-Bonaventure. *Dictionnaire étymologique de la langue françoise; où, les mots sont classés par familes.* 2 vols. Paris: Decourchant, 1829.

de Seta, Caesare. *L'Italia del Grand Tour: Da Montaigne a Goethe.* Napoli: Electa, 1996.

de Vivo, Filippo. *Information and Communication in Venice: Rethinking Early Modern Politics.* Oxford: Oxford University Press, 2007.

Delbourgo, James. *A Most Amazing Scene of Wonders: Electricity and Enlightenment in Early America.* Cambridge, MA: Harvard University Press, 2006.

Della Torre, Giovanni Maria. *Storia e fenomeni del Vesuvio.* Naples: Raimondi, 1755.

———. *Elementa Physicae.* 8 vols. Naples: n.p., 1757–69.

del Negro, Piero. "L'università." In *Storia della cultura veneta,* vol. 5, *Il Settecento,* pt. 1, edited by Girolamo Arnaldi and Manlio Pastore Stocchi, 47–76. Vicenza: Neri Pozza, 1985.

Dictionnaire de l'Académie Françoise. 3 vols. Paris: J.-B. Coignard, 1740.

Diderot, Denis, and d'Alembert, Jean le Rond, eds. *Encyclopédie; ou, Dictionnaire raisonné des sciences, des arts et des métiers.* Vol. 5. Paris: Brisson, Durand, Le Breton, et Durand, 1755.

———. *Encyclopédie; ou, Dictionnaire raisonné des sciences, des arts et des métiers.* Vol. 10. Neuchâtel: Samuel Faulche, 1765.

Di Mitri, Gino Leonardo. *Storia biomedica del tarantismo nel XVIII secolo.* Florence: Olschki, 2006.

Di Mitri, Gino Leonardo, and Bernardino Fantini, eds. *La febbre del viaggio: Il grand tour scientifico nel Regno di Napoli.* Galatina: Congedo, 2002.

Donato, Clorinda. "Between Myth and Archive, Alchemy and Science in Eighteenth-Century Naples: The Cabinet of Raimondo di Sangro, Prince of San Severo." In *Life Forms in the Thinking of the Long Eighteenth Century,* edited by Keith Michael Baker and Jenna M. Gibbs, 208–32. Toronto: University of Toronto Press, 2016.

———. *The Life and Legend of Catterina Vizzani: Sexual Identity, Science, and Sensationalism in Eighteenth-Century Italy and England.* Liverpool, UK: Liverpool University Press, 2020.

Dooley, Brendan Maurice. "Le accademie." In *Storia della cultura Veneta,* vol. 5, *Il Settecento,* pt. 1, edited by Girolamo Arnaldi and M. Pastore Stocchi, 77–90. Vicenza: Neri Pozza, 1985.

———. *Science, Politics, and Society in Eighteenth-Century Italy: The "Giornale de' letterati d'Italia" and Its World.* New York: Garland, 1991.

Doyon, André, and Lucien Liaigre. *Jacques Vaucanson: Mécanicien de Génie.* Paris: Presses Universitaires de France, 1966.

du Sommerard, Edmund. *Catalogue et description des objets d'art de l'antiquité du moyen âge et de la Renaissance.* Paris: Hôtel de Cluny, 1883.

Eamon, William. *Science and the Secrets of Nature: Books of Secrets in Medieval and Early Modern Culture.* Princeton, NJ: Princeton University Press, 1994.

Eisenstein, Elizabeth L. *The Printing Press as an Agent of Change: Communications and Cultural Transformations in Early-Modern Europe.* 2 vols. 14th ed. Cambridge: Cambridge University Press, 2009.

Elsner, Jaś, and Joan-Pau Rubiés, eds. *Voyages and Visions: Towards a Cultural History of Travel.* London: Reaktion, 1999.

Evans, Robert, John Weston, and Alexander Marr, eds. *Curiosity and Wonder from the Renaissance to the Enlightenment.* Aldershot, UK: Ashgate, 2006.

Faret, Nicolas. *L'honneste homme; ou, L'art de plaire à la cour.* Paris: Toussaints du Bray, 1630.

Farinella, Calogero. "Le traduzioni italiane della *Cyclopedia* di Ephraim Chambers." *Studi settecenteschi* 16 (1996): 97–160.

Ferrone, Vincenzo. *La nuova Atlantide e i lumi: Scienza e politica nel Piemonte di Vittorio Amedeo III*. Turin: Meynier, 1988.

Findlen, Paula. "Jokes of Nature and Jokes of Knowledge: The Playfulness of Scientific Discourse in Early Modern Europe." *Renaissance Quarterly* 43, no. 2 (1990): 292–331.

———. "Science as a Career in Enlightenment Italy: The Strategies of Laura Bassi." *Isis* 84, no. 3 (1993): 441–69.

———. "Translating the New Science: Women and the Circulation of Knowledge in Enlightenment Italy." *Configurations* 3, no. 2 (1995): 167–206.

Findlen, Paula, Wendy Wassyng Roworth, and Catherine M. Sama, eds. *Italy's Eighteenth Century: Gender and Culture in the Age of the Grand Tour*. Stanford, CA: Stanford University Press, 2009.

Fino, Lucio. *The Miracle of San Gennaro: Witness Accounts in Travel Literature from the 18th to 19th Century*. Naples: Grimaldi, 2018.

Fontenelle, Bernard le Bouvier de. "Eloge de M. Rouillé." *Histoire et mémoires de l'Académie royale des sciences* (1761): 182–88.

Füssel, Marian. "'The Charlantry of the Learned': On the Moral Economy of the Republic of Letters in Eighteenth-Century Germany." *Cultural and Social History* 3, no. 3 (2006): 287–300.

Garofalo, Silvano. *L'enciclopedismo italiano: Gianfrancesco Pivati*. Ravenna: Longo, 1980.

Gauvin, Jean-François. "Le cabinet de physique du château de Cirey et la philosophie naturelle de Mme du Châtelet et de Voltaire." In *Emilie du Châtelet: Rewriting Enlightenment Philosophy and Science*, edited by Judith P. Zinsser and Julie Chandler Hayes, 165–202. Oxford, UK: Voltaire Foundation, 2006.

———. "The Instrument That Never Was: Inventing, Manufacturing, and Branding Réaumur's Thermometer during the Enlightenment." *Annals of Science* 69, no. 4 (2012): 515–49.

Gentilcore, David. *Medical Charlatanism in Early Modern Italy*. Oxford: Oxford University Press, 2006.

Gibbon, Edward. *Viaggio in Italia*. Milan: Edizioni del Borghese, 1965.

Goethe, Johann Wolfgang von. *Letters from Switzerland and Travels in Italy*. Edited by A. J. W. Morrison. New York: Harvard Publishing, 1895.

Goodman, Dena. *The Republic of Letters: A Cultural History of the French Enlightenment*. Ithaca, NY: Cornell University Press, 1994.

Golinski, Jan. *Science as Public Culture: Chemistry and Enlightenment in Britain, 1760–1820*. Cambridge: Cambridge University Press, 1992.

Goudar, Ange. *L'espion chinois; ou, L'envoyé secret de la cour de Pekin, pour examiner l'état présent de l'Europe*. 6 vols. Cologne: n.p., 1764.

———. *L'espion françois à Londres; ou, Observations critique sur l'Angleterre et sur les Anglois*. 2 vols. London: n.p., 1778.

Grafton, Anthony. *Forgers and Critics: Creativity and Duplicity in Western Scholarship*. Princeton, NJ: Princeton University Press, 1990.

———. *What Was History? The Art of History in Early Modern Europe*. Cambridge: Cambridge University Press, 2007.

———. "Invention of Traditions and Traditions of Invention in Renaissance Europe: The Strange Case of Annius of Viterbo." In *Invention of Traditions and Traditions of Invention in Renaissance Europe: The Strange Case of Annius of Viterbo*, edited by Ann Blair and Anthony Grafton, 8–38. Philadelphia: University of Pennsylvania Press, 2010.

Guidicini, Giuseppe. *Miscellanea storico-patria bolognese*. Bologna: Monti, 1872.

Haas, Angela. "Miracles on Trial: Wonders and Their Witnesses in Eighteenth-Century France." *Proceedings of the Western Society for French History* 38 (2010): 111–28.

Habermas, Jürgen. *The Structural Transformation of the Public Sphere: An Inquiry into a Category of Bourgeois Society.* Cambridge, MA: MIT Press, 1989.

Hahn, Roger. *The Anatomy of a Scientific Institution: The Paris Academy of Sciences, 1666–1803.* Berkeley: University of California Press, 1971.

[Haller, Albrecht von]. "An Historical Account of the Wonderful Discoveries Made in Germany, etc., Concerning Electricity." *Gentleman's Magazine* 15 (1745): 193–97.

Hankins, Thomas L., and Robert J. Silverman. *Instruments and the Imagination.* Princeton, NJ: Princeton University Press, 1995.

Harris, John Raymond. *Industrial Espionage and Technology Transfer: Britain and France in the Eighteenth Century.* Aldershot, UK: Ashgate, 1998.

Haskell, Yasmin. *Loyola's Bees: Ideology and Industry in Jesuit Latin Didactic Poetry.* Oxford: Oxford University Press, 2003.

Hauc, Jean-Claude. *Ange Goudar: Un aventurier des Lumières.* Paris: Champion, 2004.

Havens, Earle A. "Babelic Confusion: Literary Forgery and the Bibliotheca Fictiva." In *Literary Forgery in Early Modern Europe, 1450–1800,* edited by Walter Stephens, Earle A. Havens, and Janet E. Gomez, 33–73. Baltimore, MD: Johns Hopkins University Press, 2018.

———. "Forgery." In *Information: A Historical Companion,* edited by Ann Blair, Paul Duguid, Anja-Silvia Goeing, and Anthony Grafton, 458–63. Princeton, NJ: Princeton University Press, 2021.

Hélie, Jérôme. *Les relations internationales dans l'Europe moderne: Conflits et équilibres européens, 1453–1789.* Paris: Colin, 2008.

Heering, Peter. "The Enlightened Microscope: Re-Enactment and Analysis of Projections with Eighteenth-Century Solar Microscopes." *British Journal for the History of Science* 41, no. 3 (2008): 345–67.

Heering, Peter, Olivia Hochadel, and David J. Rhees, eds. *Playing with Fire: Histories of the Lightning Rod.* Philadelphia: American Philosophical Society, 2009.

Heilbron, John L. *Electricity in the 17th and 18th Centuries: A Study of Early Modern Physics.* Berkeley: University of California Press, 1979.

———. *Weighing the Imponderables and Other Quantitative Science around 1800.* Berkeley: University of California Press, 1993.

Hesse, Carla, and Peter Sahlins, eds. "Mobility in French History." Special issue, *French Historical Studies* 29, no. 3 (2006).

Hilaire-Pérez, Liliane. "Cultures techniques et pratiques de l'échange, entre Lyon et le Levant: Inventions et réseaux au XVIIIe siècle," *Revue d'histoire moderne et contemporaine* 49 (2002) : 89–104.

———. "Diderot's Views on Artists' and Inventors' Rights: Invention, Imitation, and Reputation." *British Journal for the History of Science* 35, no. 2 (2002): 129–50.

———. "Les échanges techniques entre la France et l'Angleterre au XVIIIe siècle: La révolution industrielle en question." In *Les circulations internationales en Europe: Années 1680–années 1780,* edited by Pierre-Yves Beaurepaire and Pierrick Pourchasse, 197–211. Rennes: Presses Universitaires de Rennes, 2011.

———. "Inventing in a World of Guilds: Silk Fabrics in Eighteenth-Century Lyon." In *Guilds, Innovation and the European Economy, 1400–1800,* edited by S. R. Epstein and Maarten Prak, 232–63. Cambridge: Cambridge University Press, 2008.

———. *L'invention technique au siècle des Lumières.* Paris: Albin Michel, 2000.

Hilaire-Pérez, Liliane, and Catherine Verna. "Dissemination of Technical Knowledge in the Middle Ages and the Early Modern Era: New Approaches and Methodological Issues." *Technology and Culture* 47, no. 3 (2006): 536–65.

Hochadel, Oliver. "The Sale of Shocks and Sparks: Itinerant Electricians in German Enlightenment." In *Science and Spectacle in the European Enlightenment,* edited by Bernadette Bensaude-Vincent and Christine Blondel, 89–101. Aldershot, UK: Ashgate, 2008.

Hutková, Karolina. "Technology Transfers and Organization: The English East India Company and the Transfer of Piedmontese Silk Reeling Technology to Bengal, 1750s–1790s." *Enterprise and Society* 18, no. 4 (2017): 921–51.

Hutton, William. *The History of Derby.* In *English Historical Documents.* Vol. 10, *1714–1783,* edited by D. B. Horn and Mary Ransome, 458–61. Oxford: Oxford University Press, 1969.

Infelise, Mario. "Criminali e cronaca nera negli strumenti pubbllici di informazione tra '600' e '700." *Acta historiae* 15 (2007): 507–20.

———. *L'editoria veneziana nel '700.* Milan: Angeli, 2000.

———. "Enciclopedie e pubblico a Venezia a metà Settecento: G. F. Pivati e i suoi dizionari." *Studi settecenteschi* 16 (1996): 161–90.

Ionescu, Christina, and Christina Smylitopoulos. "Wonder in the Eighteenth Century / L'émerveillement au dix-huitième siècle." *Lumen: Selected Proceedings from the Canadian Society for Eighteenth-Century Studies / Lumen: Travaux choisis de la Société canadienne d'étude du dix-huitième siècle* 39 (2020): v–xlvi.

Jacob, Margaret, and Larry Stewart. *Practical Matter: Newton's Science in the Service of Industry and Empire.* Cambridge, MA: Harvard University Press, 2004.

Jallabert, Jean. *Expériences sur l'électricité avec quelques conjectures sur la cause de ses effets.* Paris: Durand et David le Jeune, 1749.

Johns, Adrian. *The Nature of the Book: Print and Knowledge in the Making.* Chicago: University of Chicago Press, 1998.

Jones, Peter. *Industrial Enlightenment: Science, Technology and Culture in Birmingham and the West Midlands, 1760–1820.* Manchester, UK: Manchester University Pres, 2008.

Jouhaud, Christian. *Mazarinades: La fronde des mots.* Paris: Aubier, 1985.

J. W. "Salutary Effect of Electricity." *Gentlemen's Magazine* 17 (1747): 238.

Kaplan, Steven L. *Les ventres de Paris: Pouvoir et approvisionnement dans la France d'Ancien Régime.* Paris: Fayard, 1988.

Kavey, Allison. *Books of Secrets: Natural Philosophy in England, 1550–1600.* Urbana: University of Illinois Press, 2007.

Kratzenstein, Christian Gottlieb. *Abhandlung von dem electricitat in der Urznenwissenschaft.* In *C. G. Kratzenstein and His Studies on Electricity during the Eighteenth Century,* edited by E. Snorrason. Odense: Odense University Press, 1974.

Kraus, Franz Xaver. *Briefe Benedicts XIV an den canonicus Pier Francesco Peggi in Bologna, 1729–1758.* Freiburg: Mohr, 1888.

Kreiser, B. Robert. *Miracles, Convulsions, and Ecclesiastical Politics in Early Eighteenth-Century Paris.* Princeton, NJ: Princeton University Press, 2015.

Krüger, Johann Gottlob. *Naturlehre.* Halle: Hemmerde, 1744.

———. *Der Weltweisheit und Artzneygelahrheit Doctors und Professors auf der Friedrichs Universität Naturlehre nebst Kupfern und vollständigem Register.* Halle: Hemmerde, 1744.

———. *Zuschrift an seine Zuhörer, worinnen er Gedanken von der Electricität mittheilt und Ihnen zugleich seine künftigen Lectionen bekannt macht.* Halle: Hemmerde, 1745.

La Vopa, Anthony J. "The Birth of Public Opinion." *Wilson Quarterly* 15, no. 1 (1991): 46–55.

———. "Herder's Publikum: Language, Print, and Sociability in Eighteenth-Century Germany." *Eighteenth-Century Studies* 29, no. 1 (1995): 5–24.

Lebeau, Christine. "Circulations internationales et savoirs d'état au XVIIIe siècle." In *Les circulations internationale en Europe: Années 1690–années 1780,* edited by Pierre-Yves Beaurepaire and Pierrick Pourchasse, 169–79. Rennes: Presses Universitaires de Rennes, 2019.

Leong, Elaine Yuen Tien, and Alisha Michelle Rankin. *Secrets and Knowledge in Medicine and Science, 1500–1800*. Aldershot, UK: Ashgate, 2011.

Lever, Maurice. *Canards sanglants: Naissance du fait divers*. Paris: Fayard, 1993.

Long, Pamela O. *Openness, Secrecy, Authorship: Technical Arts and the Culture of Knowledge from Antiquity to the Renaissance*. Baltimore, MD: Johns Hopkins University Press, 2001.

Louis, Antoine. *Observation sur l'électricite; où, L'on tâche d'expliquer son méchanisme et ses effets sur l'oeconomie animale*. Paris: Osmont et de la Guette, 1747.

Lovett, Richard. *The Subtil Medium Prov'd*. London: Hinton and Sandby, 1756.

Luynes, Charles Philippe d'Albert. *Mémoires du duc de Luynes sur la cour de Louis XV (1735–1758)*, 17 vols. Paris: Firmin-Didot Frères, 1860–65.

Lynch, Jack. *Deception and Detection in Eighteenth-Century Britain*. Aldershot, UK: Ashgate, 2008.

Lynn, Michael R. *Popular Science and Public Opinion in Eighteenth-Century France*. Manchester, UK: Manchester University Press, 2006.

Macchia, Giovanni, and Massimo Colesanti, eds. *Montesquieu: Viaggio in Italia*. Rome: Laterza, 1971.

Maçzak, Antoni. *Viaggi e viaggiatori nell'Europa moderna*. Rome: Laterza, 1992.

Maffei, Scipione. *Della formazione de' fulmini trattato del sig. marchese Scipione Maffei raccolto da varie sue lettere, in alcune delle quali si tratta anche degli'insetti rigenerantisi, e de' pesci di mare su i monti, e più a lungo dell'elettricità*. Verona: Giannalberto Tumermani nella via delle Foggie, 1747.

———. *Epistolario: 1700–1755*. 2 vols. Edited by Celestine Garibotto. Milan: Ciuffré, 1955.

Marana, Giovanni Paolo. *L'espion turc dans les cours des princes chrétiens; ou, Lettres et mémoires d'un envoyé secret de la porte dans les cours de l'Europe*. 7 vols. London, 1741–43.

Margocsy, Daniel, and Koeen Vermeir, eds. "State of Secrecy." Special issue, *British Journal for the History of Science* 45 (2012).

Mariotti, Prospero. *Lettera scritta ad una dama dal signor dottore Prospero Mariotti sopra la cagione de' fenomeni della macchina elettrica*. Perugia: Costantini e Maurizi, 1748.

Markey, Lia, ed. *Renaissance Invention: Stradanus's "Nova Reperta."* Evanston, IL: Northwestern University Press, 2020.

Mazzolari, Giuseppe Maria. *Electricorum*. Vol. 6. Rome: Salomoni, 1767.

Mazzotti, Massimo. *The World of Maria Gaetana Agnesi, Mathematician of God*. Baltimore, MD: Johns Hopkins University Press, 2007.

Mazzuchelli, Giammaria. *Gli scrittori d'Italia, cioè notizie storiche e critiche intorno alle vite e agli scritti dei letterati italiani*. 6 vols. Brescia: Giambatista Bossini, 1753–63.

McClellan, James E., III, and François Regourd. *The Colonial Machine: French Science and Overseas Expansion in the Old Regime*. Turnhout: Brepols, 2011.

Melli, Elio, ed. *Studi e inediti per il primo centenario dell'Istituto magistrale "Laura Bassi": Epistolario di Laura Bassi*. Bologna: STEB, 1960.

Melton, James Van Horn. *The Rise of the Public in Enlightenment Europe*. Cambridge: Cambridge University Press, 2001.

Ménage, Gilles. *Dictionnaire étymologique; ou, Origines de la langue françoise*. 2 vols. Paris: Jean Anisson, 1694.

Messbarger, Rebecca. *The Century of Women: Representations of Women in Eighteenth-Century Italian Public Discourse*. Toronto : University of Toronto Press, 2002.

Meyssonnier, Simone. *La balance et l'horloge: La genèse de la pensée libérale en France au XVIIIe siècle*. Paris: Verdier, 1989.

Minard, Philippe. *La fortune du colbertisme: État et industrie dans la France des Lumières*. Paris: Fayard, 1998.

Minuzzi, Sabrina. *Sul filo dei segreti: Farmacopea, libri e pratiche terapeutiche a Venezia in età moderna*. Milan: Unicopli, 2016.

Misson, Maximilien. *Nouveau Voyage d'Italie de Monsieur Misson, avec un mémoire contenant des avis utiles à ceux qui voudront faire le même voyage*. Utrecht: Guillaume vande Water et Jacques van Poolsum, 1722.

Mokyr, Joel. *The Enlightened Economy: An Economic History of Britain, 1700–1850*. New Haven, CT: Yale University Press, 2009.

Molà, Luca. *The Silk Industry of Renaissance Venice*. Baltimore, MD: Johns Hopkins University Press, 2000.

Molà, Luca, Reinhold C. Mueller, and Claudio Zanier, es. *La seta in Italia dal Medioevo al Seicento: Dal baco al drappo*. Venice: Marsilio, 2000.

Montesquieu, Charles de Secondat. *Oeuvres de Montesquieu*. 8 vols. Paris: Feret, 1827.

Montesquieu, Charles de Secondat, and Albert de Secondat Montesquieu. *Voyages de Montesquieu*. Vol. 2. Bordeaux: Gounouilhou, 1896.

Morgagni, Giambattista. *Carteggio tra Giambattista Morgagni e Francesco M. Zanotti*. Edited by Gino Rocchi. Bologna: Zanichelli, 1885.

Mortimer, Cromwell. "A Letter to Martin Folkes, Esq., President of the Royal Society, from Cromwell Mortimer, M.D., Secr. of the Same, Concerning the Natural Heat of Animals." *Philosophical Transactions of the Royal Society* 53 (1744): 473–80.

Mozzillo, Atanasio. *La frontiera del Grand Tour: Viaggi e viaggiatori nel mezzogiorno borbonico*. Naples: Liguori, 1992.

———. *Passaggio a mezzogiorno: Napoli e il sud nell'immaginario barocco e illuminista europeo*. Milan: Leonardo, 1993.

Mukerji, Chandra. *Impossible Engineering: Technology and Territoriality on the Canal du Midi*. Princeton, NJ: Princeton University Press, 2015.

Muratori, Lodovico Antonio. *Edizione nazionale del carteggio di L. A. Muratori*. Vol. 14, *Carteggio con Alessandro Chiappini*. Edited by Paolo Castignoli. Florence: Olschki, 1975.

———. *Edizione nazionale del carteggio di L. A. Muratori*. Vol. 20, *Carteggio con Pietro E. Gherardi*. Edited by Guido Pugliese. Florence: Olschki, 1982.

Naddeo, Barbara. "Cultural Capitals and Cosmopolitanism in Eighteenth-Century Italy: The Historiography of Italy on the Grand Tour." *Journal of Modern Italian Studies* 10 (2005): 183–195.

Natale, Alberto. *Gli specchi della paura: Il sensazionale e il prodigioso nella letteratura di consumo (secoli XVII–XVIII)*. Rome: Carocci, 2008.

Nollet, Jean-Antoine. *Essai sur l'électricité des corps*. Paris: Guerin, 1746.

———. "Expériences et observations en différent endroits d'Italie." *Mémoires de l'Académie royale des sciences de Paris* (1753): 444–88.

——. "Extract of a Letter from the Abbé Nollet, F. R. S., etc. to Charles, Duke of Richmond, F. R. S., Accompanying an Examination of Certain Phaenomena in Electricity, Published in Italy." *Philosophical Transactions of the Royal Society* 46 (1749–50): 368–97.

———. "Extract of the Observations Made in Italy, by the Abbé Nollet, F. R. S., on the Grotta de Cani." *Philosophical Transactions of the Royal Society* 47 (1751–52): 48–61.

———. *Leçons de physique expérimentale*. Paris: Guerin, 1743–48.

———. *Lettere intorno all'elettricità*. 2 vols. Naples: Raimondi, 1762.

———. "Observations sur quelques nouveaux phénomènes d'électricité." *Mémoires de l'Académie royale des sciences de Paris* (1746): 1–23.

———. *Programme; ou, Idée générale d'un cours de physique expérimentale avec un catalogue raisonné des instrumens qui servent aux expériences*. Paris: Le Mercier, 1738.

———. *Recherches sur les causes particulières des phénomènes électriques*. Paris: Nouvelle edition, 1754.

———. "Suite des expériences et observations en différent endroits d'Italie." *Mémoires de l'Académie royale des sciences de Paris* (1754): 54–106.

Oreskes, Naomi, and Erik M. Conway. *Merchants of Doubt: How a Handful of Scientists Obscured the Truth on Issues from Tobacco Smoke to Global Warming.* London: Bloomsbury, 2011.

Pace, Antonio. *Benjamin Franklin and Italy.* Philadelphia: American Philosophy Society, 1958.

Pancaldi, Giuliano. *Volta: Science and Culture in the Age of Enlightenment.* Princeton, NJ: Princeton University Press, 2005.

Perissa Torrini, Isabella. "Il gioco e lo svago dei veneziani: I casini." *Dal Museo Alla Città* 6 (1987): 99–123.

Perkins, Maureen. *The Reform of Time: Magic and Modernity.* London: Pluto Press, 2001.

Petrella, Giancarlo. *L'officina del geografo: La* Descrittione di tutta Italia *di Leandro Alberti e gli studi geografico-antiquari tra quattro e cinquecento.* Milan: Vita e Pensiero, 2004.

Piccolino, Marco, and Marco Bresadola. *Rane, torpedini e scintille: Galvani, Volta e l'elettricità animale.* Turin: Bollati Boringhieri, 2003.

Piot, Yann. *Jean-Antoine Nollet, artisan expérimentateur: Un discours technique au XVIIIe siècle.* Paris: Classiques Garnier, 2019.

Pigatti, Giovanni Maria. Review of *Istoria d'un sonnambulo. Novelle della repubblica letteraria* (June 1744): 177–79.

Pivati, Giovanni Francesco. *Della elettricità medica lettera del chiarissimo Signore Gio. Francesco Pivati.* Lucca: n.p., 1747.

———, ed. *Nuovo dizionario scientifico e curioso sacro-profano.* 10 vols. Venice: Benedetto Milocco, 1746–51.

———. *Riflessioni fisiche sulla medicina elettrica.* Venice: Benedetto Milocco, 1749.

Plebani, Tiziana. "Socialità, conversazioni e casini nella Venezia del secondo Settecento." In *Salotti e ruolo femminile in Italia: Tra fine Seicento e primo Novecento,* edited by Maria Luisa Betri and Elena Brambilla, 153–76. Padua: Marsilio, 2004.

Poni, Carlo. "Archéologie de la fabrique: La diffusion des moulins à soie 'alla Bolognese' dans les états Vénitiens du XVIe au XVIIIe siècle." *Annales—Histoire, sciences sociales* 27, no. 6 (1972): 1475–96.

———. "Standards, Trust, and Civil Discourse: Measuring the Thickness and Quality of Silk Thread." *History of Technology* 23 (2011): 1–16.

Porter, Roy. "Laymen, Doctors, and Medical Knowledge in the Eighteenth Century: The Evidence of the *Gentleman's Magazine.*" In *Patients and Practitioners: Lay Perceptions of Medicine in Pre-Industrial Society,* edited by Roy Porter, 283–314. Cambridge: Cambridge University Press, 1985

———, ed. *Patients and Practitioners: Lay Perceptions of Medicine in Pre-Industrial Society.* Cambridge: Cambridge University Press, 1985.

———. *Quacks: Fakers and Charlatans in Medicine.* Stroud, UK: Tempus, 2003.

———. "The Sexual Politics of James Graham." *Journal for Eighteenth-Century Studies* 5, no. 2 (1982): 199–206.

Priestley, Joseph. *History and Present State of Electricity.* London: Dodsley, 1767.

Principe, Lawrence M. *The Secrets of Alchemy.* Chicago: University of Chicago Press, 2012.

Proctor, Robert, and Londa L. Schiebinger, eds. *Agnotology: The Making and Unmaking of Ignorance.* Stanford, CA: Stanford University Press, 2008.

Prosperi, Adriano. "Otras indias: Missionari della controriforma tra contadini e selvaggi." In *Scienze, credenze occulte, livelli di cultura: Convegno internazionale di studi (Firenze, 26–30 Giugno 1980),* 205–34. Florence: Olschki, 1982.

Pyenson, Lewis, and Jean-François Gauvin, eds. *The Art of Teaching Physics: The Eighteenth-Century Demonstration Apparatus of Jean Antoine Nollet.* Sillery, Quebec: Septentrion, 2002.

Quignon, Hector. *L'abbé Nollet Physicien.* Paris: Champion, 1905.

Riskin, Jessica. *The Restless Clock: A History of the Centuries-Long Argument over What Makes Living Things Tick.* Chicago: University of Chicago Press, 2016.

———. *Science in the Age of Sensibility: The Sentimental Empiricists of the French Enlightenment.* Chicago: University of Chicago Press, 2002.

Roberts, Lissa. "Heterogeneous Purposes and the Protocols of Experiment; or, How Tracing the History of Amber Can Shed Light on Medical Electricity." In *Electric Bodies: Episodes in the History of Medical Electricity*, edited by Paola Bertucci and Giuliano Pancaldi, 17–41. Bologna: Università di Bologna, 2001.

Roche, Daniel. *Les circulations dans l'Europe moderne, XVIIe–XVIII.* Paris: Fayard, 2011.

———. *Humeurs vagabondes: De la circulation des hommes et de l'utilité des voyages.* Paris: Fayard, 2003.

———. *Le siècle des Lumières en Province: Académies et académiciens provinciaux, 1680–1789.* 2 vols. Paris: Editions de l'Ecole des Hautes Etudes en Sciences Sociales, 1989.

Roero, Silvia. "Il Gabinetto di fisica nel Settecento." In *La memoria della scienza. Musei e collezioni dell'Università di Torino*, edited by Giacomo Giacobini, 53–58. Turin: Università degli studi di Torino e Fondazione CRT, 2003.

Ronfort, Jean-Nérée. "Science and Luxury: Two Acquisitions by the J. Paul Getty Musuem." *J. Paul Getty Musuem Journal* 17 (1989): 47–82.

Rousseau, Jean-Jacques. *The Miscellaneous Works of Mr. J. J. Rousseau.* Vol. 4. London: T. Becket and P. A. De Hondt, 1767.

Rubiés, Joan-Pau. "Comparing Cultures in the Early Modern World: Hierarchies, Genealogies and the Idea of European Modernity." In *Regimes of Comparatism: Frameworks of Comparison in History, Religion and Anthropology*, edited by Renaud Gagné, Simon Goldhill, and Geoffrey Llyod, 116–72. Leiden: Brill, 2019.

———. "Instructions for Travellers: Teaching the Eye to See." *History and Anthropology* 9, nos. 2–3 (1996): 139–90.

———. "Nature and Customs in Late Medieval Ethnography: Marco Polo and John Mandeville." In *La nature comme source de la morale au Moyen Age*, edited by Maaike van der Lugt, 189–232. Florence: SISMEL-Edizioni del Galluzzo, 2014.

Rubiés, Joan-Pau, and Manel Ollé. "The Comparative History of a Genre: The Production and Circulation of Books on Travel and Ethnographies in Early Modern Europe and China." *Modern Asian Studies* 50, no. 1 (2016): 259–309.

Ruestow, Edward. "Leeuwenhoek and the Campaign against Spontaneous Generation." *Journal of the History of Biology* 17, no. 2 (1984): 225–48.

Sabel, Charles F., and Jonathan Zeitlin. "Fashion as Flexible Production: The Strategies of the Lyons Silk Merchants in the Eighteenth Century." In *Worlds of Possibilities: Flexibility and Mass Production in Western Industrialization*, edited by Carlo Poni, 37–74. Cambridge: Cambridge University Press, 1997.

Safier, Neil. *Measuring the New World: Enlightenment Science and South America.* Chicago: University of Chicago Press, 2008.

Scataglini, Marco. *Tutt'intorno Roma: Viaggio alla scoperta della campagna romana.* Tuscany: Penne e Papiri, 2008.

Schaffer, Simon. "The Consuming Flame: Electrical Showmen and Tory Mystics in the World of Goods." In *Consumption and the World of Goods*, edited by John Brewer and Roy Porter, 489–526. New York: Routledge, 1993.

———. "Experimenters' Techniques, Dyers' Hands, and the Electric Planetarium." *Isis* 88, no. 3 (1997): 456–83.

———. "Glass Works: Newton's Prisms and the Uses of Experiment." In *The Uses of Experiment: Studies in the Natural Sciences*, edited by David Gooding, Trevor Pinch, and Simon Schaffer, 67–104. Cambridge: Cambridge University Press, 1989.

———. "Machine Philosophy: Demonstration Devices in Georgian Mechanics." *Osiris* 9 (1994): 157–82.

———. "Natural Philosophy and Public Spectacle in the Eighteenth Century." *History of Science* 21, no. 1 (1983): 1–43.

———. "Self Evidence." *Critical Inquiry* 18, no. 2 (1992): 327–62.

———. "Traveling Machines and Colonial Times." *Archives de sciences sociales des religions* 187, no. 3 (2019): 171–90.

Schaffer, Simon, Lissa Roberts, Kapil Raj, James Delbourgo, and H. Otto Sibum. *The Brokered World: Go-Betweens and Global Intelligence, 1770–1820.* Sagamore Beach, MA: Watson, 2009.

Schettino, Edvige. "L'insegnamento della fisica sperimentale a Napoli nella seconda metà del Settecento." *Studi settecenteschi* 18 (1998): 367–76.

Sénelier, Jean. *Voyageurs français en Italie du Moyen Age à nos jours: Premier essai de bibliographie.* Moncalieri: Centro interuniversitario di ricerche sul viaggio in Italia, 2014.

[Sguario, Eusebio, and Xavier Wabst.] *Dell'elettricismo: O sia della forze elettriche de' corpi, svelate dalla fisica sperimentale.* Venice: Presso Giovanni Battista Recurti, 1746.

Shank, J. B. *Before Voltaire: The French Origins of "Newtonian" Mechanics, 1680–1715.* Chicago: University of Chicago Press, 2018.

Shapin, Steven. *A Social History of Truth: Civility and Science in Seventeenth-Century England.* Chicago: University of Chicago Press, 1994.

Shapin, Steven, and Simon Schaffer. *Leviathan and the Air-Pump: Hobbes, Boyle, and the Experimental Life.* Princeton, NJ: Princeton University Press, 1985.

Sibum, Otto H. "Reworking the Mechanical Value of Heat: Instruments of Precision and Gestures of Accuracy in Early Victorian England." *Studies in History and Philosophy of Science* 26, no. 1 (1995): 73–106.

Signorelli, Pietro Napoli. *Vicende della coltura nelle due Sicilie.* 5 vols. Naples: Flauto, 1784.

Smith, Pamela H. "What Is a Secret?" In *Secrets and Knowledge in Medicine and Science, 1500–1800,* edited by Elaine Leong and Alisha Rankin, 47–66. Burlington, VT: Ashgate, 2011.

Smith, Pamela H., and Paula Findlen, eds. *Merchants and Marvels: Commerce, Science, and Art in Early Modern Europe.* New York: Routledge, 2002.

Snyder, Jon R. *Dissimulation and the Culture of Secrecy in Early Modern France.* Berkeley: University of California Press, 2009.

Soll, Jacob. *The Information Master: Jean-Baptiste Colbert's Secret State Intelligence System.* Ann Arbor: University of Michigan Press, 2009.

———. "The Long and Brutal History of Fake News." *Politico,* December 18, 2016. http://politi .co/2FaV5W9.

Stafford, Barbara Maria. *Artful Science: Enlightenment Entertainment and the Eclipse of Visual Education.* Cambridge, MA: MIT Press, 1994.

Stay, Benedictus. *Philosophiae versibus traditae libri vi.* Rome: Palladis, 1747.

Spector, Céline. "The 'Lights' before the Enlightenment: The Tribunal of Reason and Public Opinion." In *Let There Be Enlightenment: The Religious and Mystical Sources of Rationality,* edited by Anton M. Matytssin and Dan Edelstein, 86–102. Baltimore, MD: Johns Hopkins University Press, 2018.

Stagl, Justin. "Ars Apodemica and Socio-Cultural Research." In *Artes Apodemicae and Early Modern Travel Culture, 1550–1700,* edited by Karl A. E. Enenkel and Jan L. de Jong, 17–27. Leiden: Brill, 2019.

Stephens, Walter, Earle A. Havens, and Janet E. Gomez, eds. *Literary Forgery in Early Modern Europe, 1450–1800.* Baltimore, MD: Johns Hopkins University Press, 2018.

Sterne, Laurence. *A Sentimental Journey through France and Italy.* London: Wilson, 1807.

Stewart, Larry R. "A Meaning for Machines: Modernity, Utility, and the Eighteenth-Century British Public." *Journal of Modern History* 70, no. 2 (1998): 259–94.

———. *The Rise of Public Science: Rhetoric, Technology, and Natural Philosophy in Newtonian Britain, 1660–1750.* Cambridge, UK: Cambridge University Press, 1992.

Sutton, Geoffrey. *Science for a Polite Society: Gender, Culture, and the Demonstration of Enlightenment.* New York: Avalon, 1995.

Sweet, Rosemary, Gerrit Verhoeven, and Sarah Goldsmith, eds. *Beyond the Grand Tour: Northern Metropolises and Early Modern Travel Behaviour.* New York: Routledge, 2017.

Tanucci, Bernardo. *Epistolario.* Vol. 2. Edited by Romano P. Coppini and Rolando Nieri. Rome: Storia e letteratura, 1980.

Terrall, Mary. *Catching Nature in the Act: Réaumur and the Practice of Natural History in the Eighteenth Century.* Chicago: University of Chicago Press, 2014.

———. *The Man Who Flattened the Earth: Maupertuis and the Sciences in the Enlightenment.* Chicago: University of Chicago Press, 2002.

Todd, Dennis. *Imagining Monsters: Miscreations of the Self in Eighteenth-Century England.* Chicago: University of Chicago Press, 1995.

Torcellan, Gianfranco. *Un économiste du XVIIIe siècle: Giammaria Ortes.* Geneva: Droz, 1969.

Torlais, Jean. *Un physicien au siècle des Lumières: L'abbé Nollet, 1700–1770.* Paris: Jonas, 1987.

Urbinati, Nadia. "Physica." In *Anatomie accademiche*, vol. 2, edited by Walter Tega, 123–54. Bologna: Il Mulino, 1987.

Vaccari, Ezio. "The Organized Traveller: Scientific Instructions for Geological Travels in Italy and Europe during the Eighteenth and Nineteenth Centuries." In *Four Centuries of Geological Travel: The Search for Knowledge on Foot, Bicycle, Sledge and Camel*, edited by Patrick N. Wyse Jackson, 7–18. London: Geological Society, 2007.

Valensise, Francesca. "Impressioni di viaggio nella Calabria Ulteriore dal diario di Dominique Vivant Denon." In *ArcHistoR* 3 (2018): 475–97.

Van Swieten, Gerhard. *Vampyrismus.* Edited by Piero Violante. Palermo: Flaccovio, 1988.

Vardi, Liana. *The Physiocrats and the World of the Enlightenment.* Cambridge: Cambridge University Press, 2012.

Vedova, Giuseppe. *Biografia degli scrittori padovani.* Vol. 2. Padova: Minerva, 1836.

Venturi, Franco. "L'Italia fuori dall'Italia." In *Storia d'Italia*, vol. 3, *Dal primo Settecento all'unità d'Italia*, 985–1481. Turin: Einaudi, 1973.

———. *Settecento riformatore: La repubblica di Venezia (1761–1797).* Vol. 2. Turin: Einaudi, 1990.

Vitrioli, Diego. *Elogio di Angela Ardinghelli, napoletana.* Naples: Nobile, 1874.

Vivian, Frances. *Il console Smith mercante e collezionista.* Vicenza: Neri Pozza, 1971.

Voltaire. *Correspondence.* 107 vols. Edited by Theodore Besterman. Geneva: Institut et musée Voltaire, 1953–65.

———. *The Worlding.*

Watson, William. "An Account of Dr. Bianchini's *Recueil d'experiences faites à Venise sur le medicine electrique.*" *Philosophical Transactions of the Royal Society* 47 (1752): 399–406.

Werrett, Simon. *Fireworks: Pyrotechnic Arts and Sciences in European History.* Chicago: University of Chicago Press, 2010.

———. *Thrifty Science: Making the Most of Materials in the History of Experiment.* Chicago: University of Chicago Press, 2019.

Wesley, John. *The Desideratum; or, Electricity Made Useful.* London: Flexney et al., 1760.

———. *Primitive Physic.* London: Trye, 1747.

West, Shearer, ed. *Italian Culture in Northern Europe in the Eighteenth Century.* Cambridge: Cambridge University Press, 1999.

Williams Elizabeth. *A Cultural History of Medical Vitalism in Enlightenment Montpellier.* Ashgate, UK: Aldershot, 2003.

Wilton, Andrew, and Ilaria Bignamini, eds. *Grand Tour: Il fascino dell'Italia nel XVIII secolo.* Milan: Skira, 1997.

Winkler, John Henry. "Novum reique medicae utile electricitatis inventum." *Philosophical Transactions of the Royal Society*, no. 486 (1748): 262–71.

Withers, Charles. *Placing the Enlightenment: Thinking Geographically about the Age of Reason.* Chicago: University of Chicago Press, 2007.

Zanetti, François. "Curing with Machines: Medical Electricity in Eighteenth-Century Paris." *Technology and Culture* 54, no. 3 (2013): 503–30.

———. *L'électricité médicale dans la France des Lumières.* Oxford, UK: Voltaire Foundation, 2017.

Zanotti, Franceso Maria. "De electricitate medica." *De Bononiensi scientiarum et artium instituti atque academiae commentarii* 3 (1755): 83–87.

[Zanotti, Francesco Maria]. *Della forza attrattiva della idee: Fragmento di un'opera scritta dal signor marchese de la Tourrì a madama la marchesa di Vincour spora l'attrazione universale.* Naples: Felica Mosca, 1767.

[Zanotti.] *Amore filosofo: In occasione della nozze solenni de' nobilissimi signori marchese Francesco Albergati e contessa Teresa Orsi.* Bologna: Dalla Volpe, 1748.

———. *Della forza de' corpi che chiamano viva libri tre.* Bologna: Pisarri e Primodì, 1752.

Zinsser, Judith P. "The Many Representations of the Marquise Du Châtelet." In *Men, Women, and the Birthing of Modern Science*, edited by Judith P. Zinsser, 48–67. DeKalb: Northern Illinois University Press, 2005.

Milton Keynes UK
Ingram Content Group UK Ltd.
UKHW010124221223
434803UK00004B/64/J